INSTRUCTOR'S TESTING MANUAL

KELLI JADE HAMMER

Broward Community College

to accompany

A SURVEY OF MATHEMATICS WITH APPLICATIONS

SEVENTH EDITION AND EXPANDED SEVENTH EDITION

Allen R. Angel
Monroe Community College

Christine D. Abbott
Monroe Community College

Dennis C. Runde
Manatee Community College

PEARSON

Addison Wesley

Boston San Francisco New York
London Toronto Sydney Tokyo Singapore Madrid
Mexico City Munich Paris Cape Town Hong Kong Montreal

Reproduced by Pearson Addison-Wesley from electronic files supplied by the author.

Copyright © 2005 Pearson Education, Inc.
Publishing as Pearson Addison-Wesley, 75 Arlington Street, Boston, MA 02116

All rights reserved. This manual may be reproduced for classroom use only. Printed in the United States of America.

ISBN   0-321-20595-2

4 5 6 BB 08 07 06

PEARSON
Addison
Wesley

TABLE OF CONTENTS

A Survey of Mathematics with Applications, 7e

Chapter 1 – Critical Thinking Skills

Form 1

In Exercises 1 and 2, use inductive reasoning to determine the next three numbers in the pattern.

1. 8, 13, 18, 23, …

2. $1, -\dfrac{1}{2}, \dfrac{1}{4}, -\dfrac{1}{8}, \ldots$

3. Pick any number, add 3 to the number, and divide the sum by 2. Multiply the quotient by 6. Subtract 9 from the product.

    a) What is the relationship between the number you started with and the final answer?

    b) Arbitrarily select some different starting numbers and repeat the process. Record the original numbers and the results.

    c) Make a conjecture about the relationship between the original numbers and the final answers.

    d) Prove, using deductive reasoning, the conjecture made in part (c).

In Exercises 4 and 5, estimate the answers.

4. $\dfrac{621,000}{0.042}$

5. $0.000214 \times 89,000,000$

6. If each square represents one square unit, estimate the area of the shaded figure.

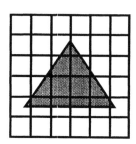

7. Below is a chart that indicates the average daily rainfall in South Florida throughout the spring and summer months.

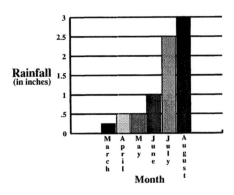

a) How much greater is the average daily rainfall in July than in May?
b) What percent of the average daily rainfall was there in May?

8. Jason has a checking account at the First Financial Bank. For this service, the bank charges a $3.50 monthly fee which includes the fee for the first ten checks written. For any checks written during the month after the first ten checks, the bank charges $0.07 per check. If Jason had to pay a total fee of $4.55 in the month of February, how many checks did he write during that month?

9. Derek needs to buy some folders for school. The Central Park Elementary school supply shop sells a 10-pack of folders for $1.25 and a single folder for $0.20. How many folders can Derek buy for $2.70?

10. If it takes Irene $6\frac{1}{4}$ hours to plant 25 geraniums, how long does it take Irene to plant each geranium?

11. Topaz, a standard poodle, can run around the block 5 times in 12 minutes. If she always runs at the same speed, how many times could she run around the block in 26.4 minutes?

12. Trevor gets paid $10 an hour if he works on a weekday and $15.50 an hour if he works during the weekend. If Trevor works a total of 17 weekday hours and 6 weekend hours, is he overpaid if his boss gives him a paycheck for $263.50? If so, by how much is Trevor being overpaid?

13. Create a magic square by using the numbers 4, 8, 12, 16, 20, 24, 28, 32, and 36. The sum of the numbers in every row, column, and diagonal must be 60.

14. Dr. Jaime Rejtman drove 80 miles to go visit his grandchildren. The first 40 miles he drove at 65 mph, and the next 40 miles he drove at 45 mph. How much time, if any, would be lost or gained if Jaime traveled the entire 80 miles at a steady speed of 55 mph?

15. From the five numbers, 5, 21, 42, 65, and 80, pick four that when added together give a total of 148.

16. At Terri's birthday party, they had a contest to guess how many hairclips were in a box. The friend that came the closest was given a prize. Three of her friends made a guess. The guesses were 16, 24, and 21. One guess was off by 6, another guess was off by 3, and another guess was off by 2. How many hairclips were in the box?

17. Larry Modica wants to purchase four lamps for her house. Sunrise Lighting has lamps on sale at two for $60. Larry has a coupon for 15% off the original price of $35 per lamp, for an unlimited number of lamps.
   a) Determine the cost of purchasing four lamps at the sale price.
   b) Determine the cost of purchasing four lamps if the coupon is used.
   c) Which is the least expensive way to purchase the four lamps, and by how much?

18. If four paintings are to be hung on a wall in a horizontal row, how many different ways can the paintings be arranged?

A Survey of Mathematics with Applications, 7e

In Exercises 1 and 2, use inductive reasoning to determine the next three numbers in the pattern.

1. $-1, 3, 7, 11, \ldots$

2. $1, \dfrac{2}{3}, \dfrac{4}{9}, \dfrac{8}{27}, \ldots$

3.   Pick any number, multiply the number by 6, and add 12 to the number. Divide the sum by 6. Subtract 2 from the quotient.
   a) What is the relationship between the number you started with and the final answer?
   b) Arbitrarily select some different starting numbers and repeat the process. Record the original numbers and the results.
   c) Make a conjecture about the relationship between the original numbers and the final answers.
   d) Prove, using deductive reasoning, the conjecture made in part (c).

In Exercises 4 and 5, estimate the answers.

4. $871{,}000 \times 0.00061$

5. $\dfrac{105{,}000}{0.00015}$

6.   If each square represents one square unit, estimate the area of the shaded figure.

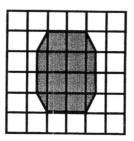

7. Below is a chart that indicates how Kerry spent her $1500 monthly paycheck.

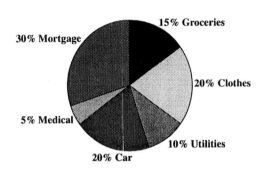

a) How much (in dollars) did Kerry spend on her mortgage?
b) How much more (in dollars) did Kerry spend on clothes than on utilities?

8. Rafael has cable television. His cable company charges $42.50 a month for basic cable which includes two premium channels. The cable company charges $3.75 a month for each additional premium channel. If Rafael's monthly cable bill is $57.50, how many premium channels does he receive?

9. At the U-Pick Place, a basket filled with 35 strawberries costs $4.89. Individually picked strawberries cost $0.15 each. What is the maximum number of strawberries that can be purchased for $15.12?

10. How much time does it take Lisa to paint 15 walls in her house if she can paint one wall in $11\frac{1}{2}$ minutes?

11. On a map, if 4 inches represents 50 miles, how many inches would represent 125 miles?

12. Steve Ray gets paid $15.50 per hour with double time for any time worked over 40 hours per week. If he works a 60-hour week and gets paid $1240, was Steve underpaid? If so, by how much?

13. Create a magic square by using the numbers 5, 7, 9, 11, 13, 15, 17, 19, 21. The sum of the numbers in every row, column, and diagonal must be 39.

14. Tamika drove 15 miles from her house to get to school. She drove the first 8 miles at 45 mph, and the next 7 miles at 30 mph. Would the trip take more, less, or the same time if Tamika drove the entire 15 miles at a steady 40 mph?

15. From the six numbers, 2, 4, 6, 8, 10, and 12, pick five that when multiplied together give a total of 4608.

16. At a local car junkyard, Brendan and his two buddies made a bet as to how many cars were piled up next to the crane. Brendan's guess was off by 6. One buddy's guess was off by 3, and his other buddy's guess was off by 5. Their three guesses were 7, 16, and 18. How many cars were piled up next to the crane?

17. Ming Lau wants to purchase 6 chairs for her dining room table. Tamarac Home Furnishings has chairs on sale at 3 for $180. Ming has a coupon for 20% off the original price of $65 per chair, for an unlimited number of chairs.
    a) Determine the cost of purchasing six chairs at the sale price.
    b) Determine the cost of purchasing six chairs if the coupon is used.
    c) Which is the least expensive way to purchase the six chairs, and by how much?

18. If eight spice jars are to be put on a kitchen shelf, in how many different ways can the spice jars be arranged?

A Survey of Mathematics with Applications, 7e

Chapter 1 – Critical Thinking Skills

In Exercises 1 and 2, use inductive reasoning to determine the next three numbers in the pattern.

1. $-4, 3, 10, 17, \ldots$

2. $1, \dfrac{1}{4}, \dfrac{1}{16}, \dfrac{1}{64}, \ldots$

3. Pick any number, subtract 6 from the number, and divide the sum by 2. Multiply the quotient by 4. Add twelve to the product.
   a) What is the relationship between the number you started with and the final answer?
   b) Arbitrarily select some different starting numbers and repeat the process. Record the original numbers and the results.
   c) Make a conjecture about the relationship between the original numbers and the final answers.
   d) Prove, using deductive reasoning, the conjecture made in part (c).

In Exercises 4 and 5, estimate the answers.

4. $\dfrac{360{,}000}{0.0016}$

5. $0.057 \times 127{,}000{,}000$

6. If each square represents one square unit, estimate the area of the shaded figure.

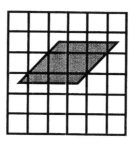

7. Below is a chart that indicates the national average price per gallon of self service regular unleaded gasoline between 1998 and 2003.

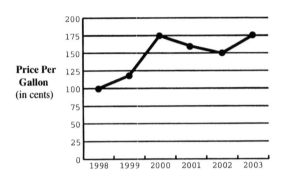

a) Which two consecutive years had the greatest increase in cost?
b) How much less did gasoline cost in 2002 than it cost in 2000?

8. Ricardo uses a cell phone company that charges $25.25 for the basic monthly fee which includes 200 minutes of airtime, plus $0.35 for each additional minute. If Ricardo's cell phone bill was $32.60 for the month of May, for how many total minutes did Ricardo use his cell phone during that month?

9. At Shop Smart Produce a bag of 10 Red Delicious apples costs $5.00 and one Red Delicious apple sold individually costs $0.72. What is the maximum number of Red Delicious apples that can be purchased for $8.60?

10. If it takes Fern $1\frac{1}{2}$ hours to learn one short dance routine, how long will it take Fern to learn 6 short dance routines?

11. Over the course of 2 days, the Lifetime TV network airs 5 reruns of The Golden Girls. How many reruns would air over the course of 14 days?

12. Robert Aranibar gets paid $12.75 per hour working the day shift and $18.50 per hour during the night shift. One week Robert works 20 daytime hours and 15 nighttime hours. If he gets a paycheck for $600 for that week, was he overpaid? If so, by how much was Robert overpaid ?

13. Create a magic square by using the numbers 9, 10, 11, 12, 13, 14, 15, 16, and 17. The sum of the numbers in every row, column, and diagonal must be 39.

14. Jodie Bigman lives 40 miles away from her sister. Driving to visit her sister one day, she drove the first 20 miles at 70 mph and the next 20 miles at 50 mph. How much time, if any, would be lost or gained if Jodie traveled the entire 40 miles at a steady speed of 60 mph?

15. From the five numbers, 6, 11, 21, 30, and 42, pick four that when added together give a total of 89.

16. The sixth grade class at John F. Kennedy middle school held a contest to see which student came closest to guessing the amount of marbles in a small jar. The three closest guesses were as follows: 58, 68, and 72. One guess was off by 8, one guess was off by 4, and the other guess was off by 6. How many marbles were in the jar?

17. Steven Agins wants to purchase nine bath towels. Linens and Things has bath towels on sale at three for $9.99. Steven has a coupon for 10% off the original price of $3.99 per towel, for an unlimited number of bath towels.

a) Determine the cost of purchasing nine towels at the sale price.
b) Determine the cost of purchasing nine towels if the coupon is used.
c) Which is the least expensive way to purchase the nine towels, and by how much?

18. If how many different ways can nine music CDs be arranged on a vertical CD rack?

A Survey of Mathematics with Applications, 7e

In Exercises 1 – 9, determine whether each is true or false. If the statement is false, explain why.

1. $\{6\} \in \{4, 5, 6, 7\}$

2. $\{2, 3\} \subset \{1, 2, 3, 4\}$

3. $\{J, S, R, K, T\}$ has twenty - five subsets.

4. $\varnothing \subseteq A$

5. For any set $A, A \cap A' = U$.

6. $2 \notin \{x \mid x \text{ is an even natural number}\}$

7. $\{c, a, t\}$ is equivalent to $\{d, o, g\}$.

8. $\{c, a, t\} = \{d, o, g\}$

9. The set of odd natural numbers less than 43 is a finite set.

In Exercises 10 and 11, use set $A = \{4, 5, 6, 7, 8\}$.

10. Write set A using set builder notation.

11. Write a description of set A.

In Exercises 12 – 15, use the following information.

$$U = \{0, 1, 2, 3, 4, 5, 6, 7, 8\}$$
$$A = \{1, 2, 3, 4, 5\}$$
$$B = \{3, 4, 5, 6, 7\}$$
$$C = \{0, 1, 7, 8\}$$

Determine the following.

12. $B \cup C$

13. $A' \cap B'$

14. $A \cap (B' \cup C)$

15. $n(B \cap C)$

16. Using the sets provided for Exercises 12 – 15, draw a Venn diagram illustrating the relationship among the sets.

17. Use a Venn diagram to determine whether $A \cup (B \cap C)' = A \cup (B' \cap C')$ for all sets A, B, and C. Show your work.

18. A novelty store sold 95 tie-dyed T-shirts. The records of their sales showed:

> 52 were green
> 51 were blue
> 48 were red
> 28 were green and blue
> 20 were blue and red
> 22 were green and red
> 8 were green, blue, and red

Construct a Venn diagram and then determine how many T-shirts sold were:

> a) only green
> b) blue and red, but no green
> c) green or red, but no blue
> d) none of these colors
> e) at least two of these colors
> f) exactly one of the colors

19. Show that the following set is infinite by setting up a one-to-one correspondence between the given set and a proper subset of itself: $\{4, 8, 12, 16, ...\}$

20. Show that the following set has cardinal number $\aleph_0$ by setting up a one-to-one correspondence between the set of counting numbers and the given set: $\{2, 5, 8, 11, ...\}$

Chapter 2 – Sets   **Form 2**

In Exercises 1 – 9, determine whether each is true or false. If the statement is false, explain why.

1. $\{7\} \in \{5, 7, 9, 11\}$

2. $\{5, 7\} \subset \{5, 7\}$

3. $\{green, blue, red\}$ has seven proper subsets.

4. $\{5\} \not\subset \{1, 2, 3, 4, 5\}$

5. $S \subset \{S, P, L, A, T\}$

6. $\{8\} \subseteq \{x \mid x \in N \text{ and } x < 8\}$

7. $\{R, O, S, E\} = \{E, S, O, R\}$

8. $\{S, I, X\}$ is equivalent to $\{6\}$.

9. $\{x \mid x \in N \text{ and } x < 12\}$ is an infinite set.

In Exercises 10 and 11, use set $S = \{1, 2, 3, 4\}$.

10. Write set S using set builder notation.

11. Write a description of set S.

In Exercises 12 – 15, use the following information.

$$U = \{2, 4, 6, 8, 10, 12\}$$
$$A = \{2, 4, 6\}$$
$$B = \{8, 10, 12\}$$
$$C = \{4, 6, 8, 10\}$$

Determine the following.

12. $A \cap B$

13. $B' \cup C$

14. $B' \cap (B \cup C)'$

15. $n(A' \cup B')$

16. Using the sets provided for Exercises 12 – 15, draw a Venn diagram illustrating the relationship among the sets.

17. Use a Venn diagram to determine whether $(A \cup B)' \cap C = (A \cap B)' \cap (A \cup B)'$ for all sets A, B, and C. Show your work.

18. For a class project, Benjamin asked 130 of his classmates what reality shows they watched. The results are as follows:

> 87 students watch American Idol
> 76 students watch Survivor
> 60 students watch The Apprentice
> 55 students watch American Idol and Survivor
> 40 students watch Survivor and The Apprentice
> 45 students watch American Idol and The Apprentice
> 35 students watch all three reality shows

Construct a Venn diagram and then determine how many students watch:

> a) exactly one of these reality shows
> b) none of these reality shows
> c) all three of these reality shows
> d) only American Idol
> e) Survivor and The Apprentice, but not American Idol
> f) at least two of these reality shows

19. Show that the following set is infinite by setting up a one-to-one correspondence between the given set and a proper subset of itself: $\{2, 8, 14, 20, ...\}$

20. Show that the following set has cardinal number $\aleph_0$ by setting up a one-to-one correspondence between the set of counting numbers and the given set: $\{2, 4, 6, 8, ...\}$

Chapter 2 – Sets   **Form 3**

In Exercises 1 – 9, determine whether each is true or false. If the statement is false, explain why.

1. $\{k\} \notin \{k, j, h\}$

2. $b \in \{a, c, e\}$

3. $\{f, o, u, r\}$ has four subsets.

4. $\{ \ \} = \varnothing$

5. $\{x, y, z\} \subset \{x, y, z\}$

6. $\{x, y, z\} \subseteq \{x, y, z\}$

7. $\{1, 2, 3\}$ is equivalent to $\{5, 6, 7\}$.

8. $\{4, 6, 8\} \subseteq \{x \mid x \in N \text{ and } 4 < x \le 10\}$

9. For any set B, $B \cup B' = U$.

In Exercises 10 and 11, use set $C = \{x \mid x \in N \text{ and } x \le 5\}$.

10. Write set C in roster format.

11. Write a description of set C.

In Exercises 12 – 15, use the following information.

$$U = \{1, 5, 10, 15, 20, 25, 30\}$$
$$A = \{1, 10, 15\}$$
$$B = \{1, 10, 20, 25\}$$
$$C = \{5, 15, 30\}$$

Determine the following.

12. $A \cup B$

13. $C \cap B'$

14. $(A \cap C') \cup B$

15. $n(A \cup B)'$

16. Using the sets provided for Exercises 12 – 15, draw a Venn diagram illustrating the relationship among the sets.

17. Use a Venn diagram to determine whether $(A \cup B') \cup C = (A' \cup B) \cup C'$ for all sets A, B, and C. Show your work.

18. The Action Sports Network surveyed 265 of its viewers to determine which sports they watch most often. The results of their survey showed:

        140 viewers watch football
        130 viewers watch baseball
        120 viewers watch hockey
         60 viewers watch football and baseball
         55 viewers watch baseball and hockey
         50 viewers watch football and hockey
         40 viewers watch all three sports

Construct a Venn diagram and then determine how many viewers watch:

    a) football and baseball, but not hockey
    b) exactly two of these sports
    c) none of these sports
    d) only hockey
    e) exactly one of these sports
    f) football and hockey, but not baseball

19. Show that the following set is infinite by setting up a one-to-one correspondence between the given set and a proper subset of itself: $\{8, 10, 12, 14, \ldots\}$

20. Show that the following set has cardinal number $\aleph_0$ by setting up a one-to-one correspondence between the set of counting numbers and the given set: $\{5, 9, 13, 17, \ldots\}$

A Survey of Mathematics with Applications, 7e

Chapter 3 – Logic

In Exercises 1 – 3, write the statement in symbolic form.

      p: Jarred plays baseball.
      q: Sierra takes dance classes.
      r: Randi enjoys painting.

1. If Jarred plays baseball then Sierra takes dance classes, or Randi enjoys painting.

2. Randi enjoys painting if and only if, Jarred plays baseball and Sierra takes dance classes.

3. Sierra takes dance classes but Randi does not enjoy painting, and Jarred plays baseball.

In Exercises 4 and 5, use p, q, and r as above to write each symbolic statement in words.

4. $(\sim r \wedge \sim q) \to \sim p$

5. $(p \wedge q) \vee \sim r$

In Exercises 6 and 7, construct a truth table for the given statement.

6. $(\sim p \vee q) \leftrightarrow r$

7. $(p \to q) \wedge \sim r$

In Exercises 8 and 9, find the truth value of the statement.

8. $8 - 2 = 4$ and $16 + 2 = 18$

9. If Florida is not in the United States or your fingers are on your feet, then you can take a picture with a camera.

In Exercises 10 and 11, given that p is true, q is false, and r is true, determine the truth value of the statement.

10. $r \leftrightarrow (p \wedge \sim q)$

11. $(p \vee q) \to (\sim r \wedge \sim q)$

12. Determine whether the pair of statements are equivalent.

$\sim(p \vee q)$, $\qquad\qquad\qquad\qquad$ $\sim p \wedge \sim q$

In Exercises 13 and 14, determine which, if any, of the three statements are equivalent.

13. a) If the Hurricanes have a good team, then they will win the national championship.
   b) If the Hurricanes win the national championship, then they have a good team.
   c) If the Hurricanes did not win the national championship, then they did not have a good team.

14. a) My pants are not blue and my shirt is not red.
   b) It is not true that my pants are blue and my shirt is red.
   c) My pants are not blue or my shirt is not red.

15. *Translate the following argument into symbolic form. Determine whether the argument is valid or invalid by comparing the argument to a recognized form or by using a truth table.*

   If I practice playing the piano, then I will become a good musician. I am a good musician. Therefore, I practice playing the piano.

16. *Use a Euler diagram to determine whether the syllogism is valid or is a fallacy.*

   All turkeys are birds.
   All birds have feathers.
   ∴ All turkeys have feathers.

For Exercises 17 and 18, write the negation of the statement.

17. No dogs have tails.

18. Some mosquitos bite.

19. Write the converse, inverse, and contrapositive of the conditional statement, "If I get paid today, then I will go shopping."

20. Is it possible for an argument to be invalid if the premises are all true? Explain your answer.

A Survey of Mathematics with Applications, 7e

In Exercises 1 – 3, write the statement in symbolic form.

> *p*: Tara loves the computer.
> *q*: Al follows the stock market.
> *r*: Jade reads magazines.

1. Jade doesn't read magazines or Tara loves the computer, and Al doesn't follow the stock market.

2. Tara loves the computer if and only if, Al doesn't follow the stock market and Jade reads magazines.

3. If Al follows the stock market then Jade doesn't read magazines, or Tara doesn't love the computer.

In Exercises 4 and 5, use p, q, and r as above to write each symbolic statement in words.

4. $\sim q \vee (\sim p \rightarrow r)$

5. $p \wedge (q \leftrightarrow \sim r)$

In Exercises 6 and 7, construct a truth table for the given statement.

6. $\sim p \vee (q \rightarrow r)$

7. $(\sim r \leftrightarrow p) \wedge q$

In Exercises 8 and 9, find the truth value of the statement.

8. If $6 \cdot 3 = 18$, then $4 + 3 = 8$.

9. If height can be measured in pounds then a dollar bill is paper money, and Hawaii is an island.

In Exercises 10 and 11, given that p is true, q is true, and r is false, determine the truth value of the statement.

10. $(p \vee \sim r) \rightarrow \sim q$

11. $(\sim p \wedge q) \leftrightarrow (\sim q \vee r)$

12. Determine whether the pair of statements are equivalent.

$p \rightarrow q$, $\sim p \wedge q$

In Exercises 13 and 14, determine which, if any, of the three statements are equivalent.

13. a) It is false that Trixi is a poodle and Kissi is a chihuahua.
     b) Trixi is not a poodle and Kissi is not a chihuahua.
     c) Trixi is not a poodle or Kissi is not a chihuahua.

14. a) If I go swimming, then the water is warm.
     b) If the water is warm, then I will go swimming.
     c) The water is not warm or I will go swimming.

15. *Translate the following argument into symbolic form. Determine whether the argument is valid or invalid by comparing the argument to a recognized form or by using a truth table.*

José repairs electronic equipment or Sam is a firefighter. José does not repair electronic equipment. Therefore, Sam is a firefighter.

16. *Use a Euler diagram to determine whether the syllogism is valid or is a fallacy.*

All dancers are graceful.
Skye is graceful.
∴ Skye is a dancer.

In Exercises 17 and 18, write the negation of the statement.

17. All disco music is great.

18. Consuelo likes computers and Jacob likes football.

19. Write the converse, inverse, and contrapositive of the conditional statement, "If Amber likes reality TV, then she will watch The Bachelorette."

20. Is it possible for an argument to be invalid if the conclusion is a true statement? Explain your answer.

A Survey of Mathematics with Applications, 7e

In Exercises 1 – 3, write the statement in symbolic form.

> *p*: Sarah is an excellent seamstress.
> *q*: Irene is a good cook.
> *r*: Max does construction.

1. Sarah is an excellent seamstress and Irene is not a good cook, or Max does not do construction.

2. If Max does construction, then Sarah is not an excellent seamstress and Irene is not a good cook.

3. Irene is a good cook if and only if Sarah is not an excellent seamstress, and Max does construction.

In Exercises 4 and 5, use p, q, and r as above to write each symbolic statement in words.

4. $(p \vee \sim q) \rightarrow r$

5. $(q \leftrightarrow \sim r) \wedge \sim p$

In Exercises 6 and 7, construct a truth table for the given statement.

6. $\sim p \vee (q \wedge r)$

7. $(\sim r \rightarrow q) \vee p$

In Exercises 8 and 9, find the truth value of the statement.

8. If $2 - 1 = 2$, then $6 - 1 = 5$.

9. A pencil can write on paper if and only if a pen can light a fire.

In Exercises 10 and 11, given that p is false, q is true, and r is true, determine the truth value of the statement.

10. $(\sim p \vee q) \wedge (\sim p \vee r)$

11. $(p \rightarrow \sim r) \vee (\sim p \rightarrow \sim r)$

12. Determine whether the pair of statements are equivalent.

   $\sim p \rightarrow \sim q$, $p \vee \sim q$

In Exercises 13 and 14, determine which, if any, of the three statements are equivalent.

13. a) The car has leather or I will buy it.
    b) If the car has leather, then I will buy it.
    c) The car does not have leather or I will buy it.

14. a) It is false that Clay is a great singer and Antonio is a great dancer.
    b) Clay is not a good singer and Antonio is not a good dancer.
    c) Clay is not a good singer or Antonio is not a good dancer.

15. *Translate the following argument into symbolic form. Determine whether the argument is valid or invalid by comparing the argument to a recognized form or by using a truth table.*

    If the new job is in Florida, then I will move. The new job is not in Florida. Therefore, I will not move.

16. *Use a Euler diagram to determine whether the syllogism is valid or is a fallacy.*

    No dogs have antlers.
    All poodles are dogs.
    ──────────────────
    ∴ No poodles have antlers.

For Exercises 17 and 18, write the negation of the statement.

17. Some cats have green eyes.

18. All babies are cute.

19. Write the converse, inverse, and contrapositive of the conditional statement, "If Suede is the pitcher, then his baseball team will win."

20. Explain how you can determine whether two statements are equivalent.

1. What is the most common type of numeration system used in the world today?

In Exercises 2 – 7, convert the numeral to a Hindu-Arabic numeral.

2. ⋂⋂౨౨౨౨‖‖‖‖‖

3. MMCMLIV

4. 九
   百
   二
   十
   八

5. $\in' \phi \delta$

6. ᛐ   ᛦᛐ   ᛦᛦᛐᛐ

7. ••••

   •

   ••

In Exercises 8 – 12, convert the number written in base 10 to a numeral in the numeration system indicated.

8. 2345 to Egyptian

9. 3643 to Roman

10. 675 to Chinese

11. 4278 to Ionic Greek

12. 6250 to Mayan

In Exercises 13 and 14, describe briefly each of the systems of numeration. Explain how each type of numeration system is used to represent numbers.

    13. Additive system

    14. Place-value system

In Exercises 15 – 18, convert the numeral to a numeral in base 10.

    15. 212_3

    16. 32_4

    17. 1011_2

    18. 765_8

In Exercises 19 – 22, convert the base 10 numeral to a numeral in the base indicated.

    19. 28 to base 2

    20. 675 to base 9

    21. 3121 to base 4

    22. 4120 to base 5

In Exercises 23 – 26, perform the indicated operations.

    23. $\begin{array}{r} 425_6 \\ +152_6 \\ \hline \end{array}$

    24. $\begin{array}{r} 352_7 \\ -213_7 \\ \hline \end{array}$

    25. $\begin{array}{r} 23_4 \\ \times 21_4 \\ \hline \end{array}$

    26. $4_5 \overline{)1431_5}$

    27. Multiply 26×35, using duplation and mediation.

    28. Multiply 31×126, using the galley method.

A Survey of Mathematics with Applications, 7e

Chapter 4 – Systems of Numeration

Form 2

1. What is a system of numeration?

In Exercises 2 – 7, convert the numeral to a Hindu-Arabic numeral.

2. MCMLXXXIV

3. 五
   千
   零
   百
   四
   十
   三

4. $\omega\xi\beta$

5. •••
   ⬯
   ••

6. < ≪▼▼ <▼

7. ∽◁𝄇99990∩∩∩∩∩

In Exercises 8 – 12, convert the number written in base 10 to a numeral in the numeration system indicated.

8. 3025 to Roman

9. 870 to Chinese

10. 4255 to Ionic Greek

11. 6821 to Mayan

12. 3712 to Babylonian

In Exercises 13 and 14, describe briefly each of the systems of numeration. Explain how each type of numeration system is used to represent numbers.

   13. Ciphered system

   14. Multiplicative system

In Exercises 15 – 18, convert the numeral to a numeral in base 10.

   15. 214_5

   16. 612_7

   17. 10101_2

   18. 541_6

In Exercises 19 – 22, convert the base 10 numeral to a numeral in the base indicated.

   19. 536 to base 3

   20. 6215 to base 8

   21. 320 to base 2

   22. 3200 to base 7

In Exercises 23 – 26, perform the indicated operations.

   23. $\begin{array}{r} 210_3 \\ +122_3 \\ \hline \end{array}$

   24. $\begin{array}{r} 1110_2 \\ -1001_2 \\ \hline \end{array}$

   25. $\begin{array}{r} 43_5 \\ \times 12_5 \\ \hline \end{array}$

   26. $2_4 \overline{)1321_4}$

   27. Multiply 38×40, using duplation and mediation.

   28. Multiply 26×305, using the galley method.

A Survey of Mathematics with Applications, 7e

Chapter 4 – Systems of Numeration Form 3

1. Why was the Babylonian system not a true place-value system?

In Exercises 2 – 7, convert the numeral to a Hindu-Arabic numeral.

2. $\gamma'\tau\lambda\gamma$

3. ⅠⅠⅠ ⟪Ⅰ ⟨

4. 四
   千
   六
   百
   七
   十
   五

5. ..
   ...
   .
   ≡

6. ⌇999ⅠⅠⅠ

7. MMMCMIV

In Exercises 8 – 12, convert the number written in base 10 to a numeral in the numeration system indicated.

8. 1838 to Ionic Greek

9. 2729 to Babylonian

10. 5466 to Chinese

11. 1224 to Egyptian

12. 712 to Mayan

In Exercises 13 and 14, describe briefly each of the systems of numeration. Explain how each type of numeration system is used to represent numbers.

13. Place-value system

14. Ciphered system

In Exercises 15 – 18, convert the numeral to a numeral in base 10.

15. 11110_2

16. 816_9

17. 1332_4

18. 545_6

In Exercises 19 – 22, convert the base 10 numeral to a numeral in the base indicated.

19. 1327 to base 6

20. 812 to base 2

21. 9171 to base 5

22. 1214 to base 4

In Exercises 23 – 26, perform the indicated operations.

23. $\begin{array}{r} 11011_2 \\ +1101_2 \\ \hline \end{array}$

24. $\begin{array}{r} 387_9 \\ -128_9 \\ \hline \end{array}$

25. $\begin{array}{r} 234_5 \\ \times 21_5 \\ \hline \end{array}$

26. $6_8 \overline{)2514_8}$

27. Multiply 31×75, using duplation and mediation.

28. Multiply 245×15, using the galley method.

Chapter 5 – Number Theory and the Real Number System   Form 1

1. Which of the numbers 2, 3, 4, 5, 6, 8, 9, and 10 divide 34,320?

2. Find the prime factorization of 744.

3. Evaluate $[-5+8]+(-6)$.

4. Evaluate $-10-(-5)$.

5. Evaluate $[45(-6)]\div(-2-3)$.

6. Convert $3\dfrac{7}{11}$ to an improper fraction.

7. Convert $\dfrac{123}{6}$ to a mixed number.

8. Write $\dfrac{1}{9}$ as a terminating or repeating decimal number.

9. Express 2.25 as a quotient of two integers.

10. Evaluate $\left(4\div\dfrac{5}{6}\right)+\left(\dfrac{3}{8}\cdot\dfrac{4}{11}\right)$.

11. Perform the operation and reduce the answer to lowest terms: $\dfrac{5}{24}-\dfrac{1}{16}$.

12. Simplify $\sqrt{40}+\sqrt{90}$.

13. Rationalize the denominator $\dfrac{6}{\sqrt{2}}$.

14. Determine whether the natural numbers are closed under the operation of subtraction. Explain your answer.

In Exercises 15 and 16, name the property illustrated.

15. $(x+y)+z=z+(x+y)$

16. $2\cdot(3\cdot4)=(2\cdot3)\cdot4$

17. Evaluate $7^2 \cdot 7^{-5}$.

18. Evaluate $\dfrac{4^{10}}{4^5}$.

19. Evaluate $\left(3^2\right)^4$.

20. Perform the operation by first converting the numerator and denominator to scientific notation. Write the answer in scientific notation. $\dfrac{0.000006}{50,000}$

21. Write an expression for the general or nth term, a_n, of the sequence $-6, -2, 2, 6, \ldots$

22. Find the sum of the terms of the arithmetic sequence. The number of terms, n, is given. $-10, -8, -6, -4, \ldots, 18;\ n = 13$

23. Find a_7 when $a_1 = 5$ and $r = 4$.

24. Find the sum of the first seven terms of the sequence when $a_1 = 7$ and $r = 2$.

25. Write an expression for the general or nth term, a_n, of the sequence $5, -10, 20, -40, \ldots$

26. Determine whether the following sequence is a Fibonacci-type sequence. If so, determine the next two terms of the sequence. $-6, 4, -2, 2, 0, 2, \ldots$

A Survey of Mathematics with Applications, 7e

Chapter 5 – Number Theory and the Real Number System   Form 2

1. Which of the numbers 2, 3, 4, 5, 6, 8, 9, and 10 divide 68,040?

2. Find the prime factorization of 680.

3. Evaluate $[-3-(-3)]-10$.

4. Evaluate $10-18$.

5. Evaluate $[14(-3)]\div(-2-5)$.

6. Convert $6\dfrac{4}{11}$ to an improper fraction.

7. Convert $\dfrac{121}{6}$ to a mixed number.

8. Write $\dfrac{3}{11}$ as a terminating or repeating decimal number.

9. Express 3.15 as a quotient of two integers.

10. Evaluate $\left(\dfrac{6}{7}\cdot\dfrac{2}{3}\right)-\left(\dfrac{7}{9}\div 2\right)$.

11. Perform the operation and reduce the answer to lowest terms: $\dfrac{9}{20}+\dfrac{11}{30}$.

12. Simplify $\sqrt{125}-\sqrt{80}$.

13. Rationalize the denominator $\dfrac{\sqrt{3}}{\sqrt{5}}$.

14. Determine whether whole numbers are closed under the operation of subtraction. Explain your answer.

In Exercises 15 and 16, name the property illustrated.

15. $a(b+c)=ab+ac$

16. $11\cdot(1\cdot 5)=(1\cdot 5)\cdot 11$

17. Evaluate $\left(\dfrac{2}{3}\right)^4$.

18. Evaluate $\dfrac{6^2}{6^4}$.

19. Evaluate $7^4 \cdot 7^{-1}$.

20. Perform the operation by first converting the numerator and denominator to scientific notation. Write the answer in scientific notation.
$(460{,}000{,}000)(0.00062)$

21. Write an expression for the general or nth term, a_n, of the sequence $-3, 1, 5, 9, \ldots$

22. Find the sum of the terms of the arithmetic sequence. The number of terms, n, is given. $-4, -9, -14, -19, \ldots, -49$; $n = 10$

23. Find a_7 when $a_1 = 6$ and $r = \dfrac{1}{2}$.

24. Find the sum of the first seven terms of the sequence when $a_1 = 3$ and $r = -2$.

25. Write an expression for the general or nth term, a_n, of the sequence $-2, 6, -18, 54, \ldots$

26. Determine whether the following sequence is a Fibonacci-type sequence. If so, determine the next two terms of the sequence. $-2, 5, 3, 9, 13, 23, \ldots$

A Survey of Mathematics with Applications, 7e

Chapter 5 – Number Theory and the Real Number System   Form 3

1. Which of the numbers 2, 3, 4, 5, 6, 8, 9, and 10 divide 42,360?

2. Find the prime factorization of 900.

3. Evaluate $[-4+10]+(-3)$.

4. Evaluate $-10-(-16)$.

5. Evaluate $[(-7)(-24)] \div (-3-3)$.

6. Convert $5\dfrac{1}{6}$ to an improper fraction.

7. Convert $\dfrac{125}{10}$ to a mixed number.

8. Write $\dfrac{1}{8}$ as a terminating or repeating decimal number.

9. Express 3.55 as a quotient of two integers.

10. Evaluate $\left(5 \div \dfrac{15}{16}\right) + \left(\dfrac{11}{12} \cdot \dfrac{5}{33}\right)$.

11. Perform the operation and reduce the answer to lowest terms: $\dfrac{13}{36} - \dfrac{7}{24}$.

12. Simplify $\sqrt{128} - \sqrt{98}$.

13. Rationalize the denominator $\dfrac{5}{\sqrt{10}}$.

14. Determine whether the whole numbers are closed under the operation of addition. Explain your answer.

In Exercises 15 and 16, name the property illustrated.

15. $5 + (x+6) = (5+x)+6$

16. $3 \cdot b = b \cdot 3$

17. Evaluate $\left(2^3\right)^2$.

18. Evaluate $6^4 \cdot 6^{-6}$.

19. Evaluate $\dfrac{11^{10}}{11}$.

20. Perform the operation by first converting the numerator and denominator to scientific notation. Write the answer in scientific notation. $\dfrac{6{,}800{,}000}{0.000017}$

21. Write an expression for the general or nth term, a_n, of the sequence $-11, -15, -19, -23, \ldots$

22. Find the sum of the terms of the arithmetic sequence. The number of terms, n, is given. $98, 95, 92, 89, \ldots, 65; \; n = 12$

23. Find a_6 when $a_1 = -2$ and $r = 4$.

24. Find the sum of the first three terms of the sequence when $a_1 = 4$ and $r = \dfrac{1}{3}$.

25. Write an expression for the general or nth term, a_n, of the sequence $12, 6, 3, \dfrac{3}{2}, \ldots$

26. Determine whether the following sequence is a Fibonacci-type sequence. If so, determine the next two terms of the sequence. $3, -3, 0, -3, -3, -6, \ldots$

Chapter 6 – Algebra, Graphs, and Functions **Form 1**

1. Evaluate $2x^2 + 6x + 5$, when $x = -3$.

In Exercises 2 and 3, solve the equation.

2. $3(4x + 2) + 5x = 30 - x$

3. $4 - 5(x + 2) = 3(x + 1) - 1$

In Exercises 4 and 5, write an equation to represent the problem. Then solve the equation.

4. Five less than three times a number is 10.

5. The cost of a CD player including 6% sales tax is $344.50. Determine the cost of the CD player before tax.

6. Evaluate $V = \dfrac{1}{3}bh$ when $b = 9$ and $h = 7$.

7. Solve $2x - 6y = 10$ for y.

8. r varies directly as p and inversely as q. If $r = 2$ when $p = 5$ and $q = 25$, find r when $p = 6$ and $q = 2$.

9. For a constant area, the base, b, of a triangle varies inversely as the height, h. If the base is 6 feet when the height is 11 feet, find the base of a triangle with the same area if the height is 3 feet.

10. Graph the solution set of $-2x - 6 > 3x + 4$ on the real number line.

11. Determine the slope of the line through the points $(4, -1)$ and $(-2, 6)$.

In Exercises 12 and 13, graph the equation.

12. $y = -3x + 1$

13. $4x + 5y = 8$

14. Graph the inequality $2x + 3y \le 6$.

15. Solve the equation $x^2 = x + 56$ by factoring.

16. Solve the equation $4x^2 + x = 7$ by using the quadratic formula.

17. Determine whether the graph is a function. Explain your answer.

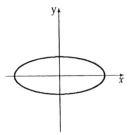

18. Evaluate $f(x) = -6x^2 - 2x + 8$ when $x = -3$.

19. For the equation $y = x^2 - 4x + 3$,
   a) determine whether the parabola will open upward or downward.
   b) determine the equation of the axis of symmetry.
   c) determine the vertex.
   d) determine the y-intercept.
   e) determine the x-intercepts if they exist.
   f) sketch the graph.
   g) determine the domain and range of the function.

A Survey of Mathematics with Applications, 7e

Chapter 6 – Algebra, Graphs, and Functions

Form 2

1. Evaluate $-3x^4 - 6x - 2$, when $x = 2$.

In Exercises 2 and 3, solve the equation.

2. $3(2x - 7) = 4x - 6x + 5$

3. $-6(x + 7) + 5 = 2(x - 3) + 3x$

In Exercises 4 and 5, write an equation to represent the problem. Then solve the equation.

4. The sum of a number and 6 is -12.

5. After a 20% discount, the final price of a summer dress is $25.60. Determine the original price of the dress before the discount was taken.

6. Evaluate $S = R - rR$ when $R = 45$ and $r = 0.3$.

7. Solve $4x - 5y = 20$ for y.

8. x varies directly as y and inversely as the square of z. If $x = 10$ when $y = 8$ and $z = 2$, find x when $y = 12$ and $z = 3$.

9. The amount of interest earned on an investment, I, varies directly as the interest rate, r. If the interest earned is $200 when the interest rate is 3.5%, find the amount of interest earned when the rate is 5%.

10. Graph the solution set of $-2x + 16 > 3x - 14$ on the real number line.

11. Determine the slope of the line through the points $(0, -5)$ and $(-10, 10)$.

In Exercises 12 and 13, graph the equation.

12. $y = \dfrac{1}{2}x + 4$

13. $3x - 5y = 9$

14. Graph the inequality $4x - y \geq 8$.

15. Solve the equation $2x^2 - 7x = 15$ by factoring.

16. Solve the equation $4x^2 + x = 6$ by using the quadratic formula.

17. Determine whether the graph is a function. Explain your answer.

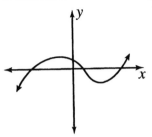

18. Evaluate $f(x) = -6x^2 + x - 5$ when $x = -3$.

19. For the equation $y = -x^2 + 4$,
   a) determine whether the parabola will open upward or downward.
   b) determine the equation of the axis of symmetry.
   c) determine the vertex.
   d) determine the y-intercept.
   e) determine the x-intercepts if they exist.
   f) sketch the graph.
   g) determine the domain and range of the function.

A Survey of Mathematics with Applications, 7e

Chapter 6 – Algebra, Graphs, and Functions   Form 3

1. Evaluate $2x^3 - x^2 - x$, when $x = 4$.

In Exercises 2 and 3, solve the equation.

2. $-4x - 5 = -2(3x + 2)$

3. $2(x - 4) + 4 = x + 3x - 12$

In Exercises 4 and 5, write an equation to represent the problem. Then solve the equation.

4. Seven less than the product of a number and 2 is 21.

5. The length of a rectangle is 3 feet longer than its width. If the perimeter of the rectangle is 103 feet, find the length and width of the rectangle

6. Evaluate $A = P(1 + rt)$ when $P = 3000$, $r = 0.02$, and $t = 6$.

7. Solve $2x + 5y = 15$ for y.

8. Q varies jointly as R and S and inversely as T. If $Q = 6$ when $R = 2$, $S = 8$, and $T = 3$, find Q when $R = 40$, $S = 16$, and $T = 2$.

9. For a constant distance the rate of travel, r, varies inversely as the time traveled, t. If the rate traveled is 60 mph when the time traveled is 2 hours, find the time traveled when the rate is 40 mph.

10. Graph the solution set of $-x - 1 < 3x + 7$ on the real number line.

11. Determine the slope of the line through the points $(3, -2)$ and $(5, 6)$.

In Exercises 12 and 13, graph the equation.

12. $y = -\dfrac{1}{2}x + 3$

13. $6x - 3y = 9$

14. Graph the inequality $4y < -3x + 8$.

15. Solve the equation $x^2 = -x + 20$ by factoring.

16. Solve the equation $2x^2 + 3x = 7$ by using the quadratic formula.

17. Determine whether the graph is a function. Explain your answer.

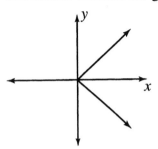

18. Evaluate $f(x) = -x^2 + 5x - 4$ when $x = 6$.

19. For the equation $y = x^2 - 6x + 8$,
   a) determine whether the parabola will open upward or downward.
   b) determine the equation of the axis of symmetry.
   c) determine the vertex.
   d) determine the y-intercept.
   e) determine the x-intercepts if they exist.
   f) sketch the graph.
   g) determine the domain and range of the function.

A Survey of Mathematics with Applications, 7e

Chapter 7 – Systems of Linear Equations and Inequalities   Form 1

1. If a system of linear equations has no solution, describe what the graph of the linear system would look like.

2. Solve the following system of equations graphically.

$$y = 3x - 1$$
$$x - 3y = 3$$

3. Determine without graphing whether the system of equations has exactly one solution, no solution, or an infinite number of solutions.

$$2x - 3y = 6$$
$$-4x + 6y = -12$$

In Exercises 4 – 9, solve the system of equations by the method indicated.

4. $2x + y = 9$
   $y = 2x - 3$
   (substitution)

5. $4x - 3y = 3$
   $6x - 2y = 2$
   (substitution)

6. $3x - 4y = 5$
   $x - 8y = 10$
   (addition)

7. $3x + 2y = 27$
   $6x - 6y = 24$
   (addition)

8. $4x + 5y = 5$
   $6x + 9y = 3$
   (addition)

9. $x + y = 1$
   $7x + 4y = 13$
   (matrices)

In Exercises 10 – 12, for $A = \begin{bmatrix} -5 & 4 \\ 6 & 1 \end{bmatrix}$ and $B = \begin{bmatrix} 2 & -3 \\ -5 & -2 \end{bmatrix}$, determine the following.

10. $A - B$

11. $4A + B$

12. $A \times B$

13. Graph the following system of linear inequalities and indicate the solution set.

$$y \geq 4x - 2$$
$$y < -3x + 1$$

In Exercises 14 and 15, solve by using a system of linear equations.

14. Jake wants to buy some dark chocolate and some milk chocolate and mix them together. The dark chocolate sells for $3.00 per pound and the milk chocolate sells for $2.50 per pound. He wants to buy a total of 15 pounds of chocolate that sells for $2.67 per pound. How many pounds of each type of chocolate should Jake buy?

15. The Simpson family could rent a van from U-Haul-It for $25.75 a day plus $0.25 per mile. If they rented a van from Ride-Around-Town, it would cost $18.00 a day plus $0.50 per mile.

a) How many miles would the Simpson family have to drive the van for the total charge to be the same from both companies?

b) If the Simpson family needs to drive 40 miles, which rental company would be more expensive?

16. The set of constraints and profit formula for a linear programming problem are as follows.

$$6x + 3y \leq 12$$
$$2x + 5y \leq 10$$
$$x \geq 0$$
$$y \geq 0$$
$$P = 10x + 16y$$

a) Draw the graph of the constraints and determine the vertices of the feasible region.

b) Use the vertices to determine the maximum and minimum profit.

A Survey of Mathematics with Applications, 7e

Chapter 7 – Systems of Linear Equations and Inequalities   Form 2

1. If a linear system has infinitely many solutions, describe what the graph of the linear system would look like.

2. Solve the following system of equations graphically.

$$2x - y = 2$$
$$y = -4x + 10$$

3. Determine without graphing whether the system of equations has exactly one solution, no solution, or an infinite number of solutions.

$$x - y = 3$$
$$3x - 3y = 12$$

In Exercises 4 – 9, solve the system of equations by the method indicated.

4. $x = 2y + 5$
   $2x + 3y = 3$
   (substitution)

5. $x - 2y = 2$
   $3x - 3y = -3$
   (substitution)

6. $6x - 2y = 10$
   $2x + y = 10$
   (addition)

7. $8x + 2y = 10$
   $2x - 3y = 13$
   (addition)

8. $2x - 3y = 2$
   $5x + 4y = 51$
   (addition)

9. $x - y = 3$
   $2x + y = 3$
   (matrices)

In Exercises 10 – 12, for $A = \begin{bmatrix} 3 & 2 \\ -1 & 2 \end{bmatrix}$ and $B = \begin{bmatrix} -6 & 1 \\ 0 & -4 \end{bmatrix}$, determine the following.

10. $A + B$

11. $A - 3B$

12. $A \times B$

13. Graph the following system of linear inequalities and indicate the solution set.

$$y < x + 5$$
$$y \geq 2x + 3$$

In Exercises 14 and 15, solve by using a system of linear equations.

14. The organization, Parents Without Partners, held an event at a local carnival. The cost of a child's ticket was $3. The cost of an adult's ticket was $5. If the organization bought 25 tickets for an average cost of $3.75 per ticket, how many of each type of ticket was purchased?

15. A new-car salesman receives a base salary of $500 weekly and $20 for every car he sells. A used-car salesman receives a base salary of $600 weekly and $15 for every car he sells.

a) How many cars would each salesman have to sell in a week for their earnings to be equal to each other?

b) If Katherine Adelstein of Pompano Honda could sell ten cars in one week, would she make higher total salary selling new cars or used cars?

16. The set of constraints and profit formula for a linear programming problem are as follows.
$$2x + y \leq 8$$
$$x + y \geq 4$$
$$x \geq 0$$
$$y \geq 0$$
$$P = 3x + 2y$$

a) Draw the graph of the constraints and determine the vertices of the feasible region.

b) Use the vertices to determine the maximum and minimum profit.

Chapter 7 – Systems of Linear Equations and Inequalities   Form 3

1. If a system of linear equations has one solution, describe what the graph of the linear system would look like.

2. Solve the following system of equations graphically.

$$y = 3x - 6$$
$$x + y = 6$$

3. Determine without graphing whether the system of equations has exactly one solution, no solution, or an infinite number of solutions.

$$x + y = -1$$
$$3x - 3y = 3$$

In Exercises 4 – 9, solve the system of equations by the method indicated.

4. $y = -2x + 4$
   $14x - 4y = 6$
   (substitution)

5. $x + 3y = 2$
   $x - y = 2$
   (substitution)

6. $2x + y = 5$
   $5x + 2y = 8$
   (addition)

7. $2x - 2y = 6$
   $3x - 4y = -3$
   (addition)

8. $3x + 6y = 27$
   $4x - 8y = -12$
   (addition)

9. $x - 2y = -4$
   $2x + y = 7$
   (matrices)

In Exercises 10 – 12, for $A = \begin{bmatrix} -3 & 2 \\ 1 & 0 \end{bmatrix}$ and $B = \begin{bmatrix} -7 & 6 \\ -6 & -2 \end{bmatrix}$, determine the following.

10. $A + B$

11. $A \times B$

12. $2A + 3B$

13. Graph the following system of linear inequalities and indicate the solution set.

$$y \leq -2x + 1$$
$$y \leq 5x - 2$$

In Exercises 14 and 15, solve by using a system of linear equations.

14. A car mechanic needs to mix together two solutions of anti-freeze. One solution is 100% pure anti-freeze and the other solution is 30% anti-freeze. The mechanic would like to have 2.5 gallons of a mixture that is 44% anti-freeze. How many gallons of each solution will the car mechanic need?

15. At the Town and Country Nursery, Colleen can purchase a box of potted plants for $6 plus $1.25 for each bag of plant food purchased. At the Flowers and Things Nursery, she can purchase a box of potted plants for $4.50 plus $1.75 for each bag of plant food.

a) Colleen buys one box of potted plants and some bags of plant food at each nursery. How many bags of plant food must she buy at each nursery in order for her total cost to be the same at both nurseries?

b) If Colleen decides to buy one box of potted plants and five bags of plant food, which nursery would be the least expensive?

16. The set of constraints and profit formula for a linear programming problem are as follows.

$$2x + y \leq 10$$
$$2x + 3y \leq 18$$
$$x \geq 0$$
$$y \geq 0$$
$$P = 3x + 2y$$

a) Draw the graph of the constraints and determine the vertices of the feasible region.

b) Use the vertices to determine the maximum and minimum profit.

Chapter 8 – The Metric System   **Form 1**

1.   Change 165 mm to m.

2.   Change 225 dag to dg.

3.   How many times greater is a hectoliter than a deciliter?

4.   If one orange weighs approximately 200 grams, how many kilograms would a sack of twenty oranges weigh?

In Exercises 5 – 9, choose the best answer.

5.   The length of a pencil is about:
    a) 70 cm
    b) 14 cm
    c) 3 cm

6.   The surface area of a notebook cover is about:
    a) 56 cm^2
    b) 56 m^2
    c) 560 cm^2

7.   The amount of water a pool can hold is about:
    a) 49 kℓ
    b) 49 ℓ
    c) 9 kℓ

8.   The mass of a coffee mug is about:
    a) 3 kg
    b) 500 g
    c) 0.02 t

9.   The outside temperature on a sunny summer day in California is about:
    a) 88° C
    b) 55° C
    c) 31° C

10. How many times greater is a square kilometer than a square meter?

11. How many times greater is a cubic centimeter than a cubic millimeter?

12. Convert 110 lb to kilograms.

13. Convert 625 cm to inches.

14. Change 20° F to degrees Celsius.

15. Change −5° C to degrees Fahrenheit.

16. A street light may be 10 ft tall. How many centimeters is this?

17. Carlotta's kitchen sink is 50 cm long, 30 cm wide, and 40 cm deep.
  a) Determine the volume of the sink in cubic centimeters.
  b) Determine the number of liters of water the sink could hold.
  c) Determine the weight of the water in kilograms.

18. It takes 3 rolls of wallpaper to cover 5 m^2 of wall surface. Each roll of wallpaper costs $4.25. What will be the cost of the wallpaper for four walls of a bedroom 15 m long, 10 m wide, and 8 m high?

A Survey of Mathematics with Applications, 7e

Chapter 8 – The Metric System   **Form 2**

1. Change 35 hg to dg.

2. Change 128 cℓ to ℓ.

3. How many times greater is a centigram than a milligram?

4. Three men are approximately the same height of 180 centimeters each. What is the total of their combined height expressed in meters?

In Exercises 5 – 9, choose the best answer.

5. The length of a small coffee table is about:
   a) 1 m
   b) 20 cm
   c) 5 m

6. The surface area of a videotape is about:
   a) 210 mm^2
   b) 210 cm^2
   c) 2 m^2

7. The amount of water a bathtub can hold is about:
   a) 48 kℓ
   b) 480 ℓ
   c) 48 ℓ

8. The mass of a personal CD player is about:
   a) 140 kg
   b) 0.14 kg
   c) 600 g

9. The temperature of boiling water is:
   a) 212° C
   b) 450° C
   c) 100° C

10. How many times greater is a square dekameter than a square centimeter?

11. How many times greater is a cubic hectometer than a cubic decimeter?

12. Convert 65 yd to meters.

13. Convert 510 cm to feet.

14. Change 316° F to degrees Celsius.

15. Change 2° C to degrees Fahrenheit.

16. A doorway may be 8 ft tall. How many meters is this?

17. Al's home fish tank is 36 cm long, 26 cm wide, and 30 cm deep.
    a) Determine the volume of the fish tank in cubic centimeters.
    b) Determine the number of liters of water the fish tank could hold.
    c) Determine the weight of the water in kilograms.

18. It takes 2 rolls of carpet to cover a 24 m^2 of floor space. Each roll of carpeting costs $12.50. What will the cost be to carpet an area that is 10 m long by 12 m wide?

Chapter 8 – The Metric System   Form 3

1. Change 56 cm to dam.

2. Change 43 hℓ to daℓ.

3. How many times greater is a meter than a millimeter?

4. If one coffee mug can hold approximately 0.5 liters of coffee, how many liters of coffee would 20 coffee mugs hold?

In Exercises 5 – 9, choose the best answer.

5. The length of a scientific calculator is about:
   a) 40 cm
   b) 140 cm
   c) 140 mm

6. The surface area of a big screen TV is about:
   a) 5000 cm^2
   b) 1.25 m^2
   c) 8 m^2

7. The amount of soup an average sized bowl can hold is about:
   a) 24 cℓ
   b) 700 mℓ
   c) 700 cℓ

8. The mass of your math textbook is about:
   a) 90 g
   b) 25 g
   c) 2.5 kg

9. The outside temperature on an autumn day in New York is about:
   a) 44° C
   b) 4° C
   c) −5° C

10. How many times greater is a square hectometer than a square decimeter?

11. How many times greater is a cubic kilometer than a cubic centimeter?

12. Convert 6 yd to meters.

13. Convert 826 g to ounces.

14. Change $-6°$ F to degrees Celsius.

15. Change $12°$ C to degrees Fahrenheit.

16. A vaulted ceiling may be 12 feet high. How many yards is this?

17. A child's sandbox is 120 cm long, 80 cm wide, and 25 cm deep.
   a) Determine the volume of the sandbox in cubic centimeters.
   b) Determine the number of liters of sand the sandbox could hold.
   c) Determine the weight of the sand in kilograms.

18. It takes 3 rolls of drapery fabric to cover 8 m^2 of a sliding glass door. Each roll of fabric costs $8.50. What will the cost be to cover with draperies, 2 extra large sliding glass doors that measure 4 m long and 4 m high each?

A Survey of Mathematics with Applications, 7e

Chapter 9 – Geometry   **Form 1**

In Exercises 1 – 4, use the following figure to describe the set of points.

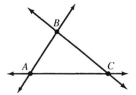

1. $\overset{\circ}{AC} \cap \overset{\circ}{CA}$

2. $\overrightarrow{BA} \cup \overrightarrow{BC}$

3. $\angle ACB \cap \angle ABC$

4. $\overset{\circ}{BC} \cap \overset{\circ}{AC}$

5. $m \angle D = 42.8°$. Determine the measure of the complement of $\angle D$.

6. $m \angle D = 78.2°$. Determine the measure of the supplement of $\angle D$.

7. Determine the measure of $\angle x$ in the following figure.

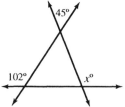

8. Determine the sum of the measures of the interior angles of a Nonagon.

9. Parallelograms $ABCD$ and $A'B'C'D'$ are similar figures. Determine the length of side $\overline{A'B'}$.

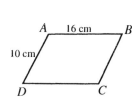 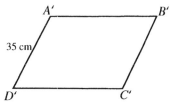

10. Right triangle ABC has one leg of length 12 cm and another leg of length 16 cm.
     a) Determine the length of the hypotenuse.
     b) Determine the perimeter of the triangle.
     c) Determine the area of the triangle.

11. Determine the volume of a right circular cylinder having a diameter of 10 ft and a height of 8 ft.

12. How much sand will a rectangular tank hold if the tank is 70 cm long, 35 cm wide, and 25 cm high?

13. Determine the volume of a sphere with a radius of 6 inches.

14. Construct a reflection of rectangle *ABCD* shown below, about line *l*. Show the rectangle in the positions both before and after reflection.

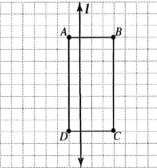

15. Construct a translation of the triangle *ABC* shown below using translation vector **v**. Show the triangle in positions both before and after the translation.

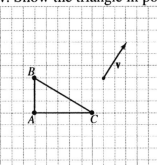

16. Construct a 90° rotation of triangle *ABC* shown below, about rotation point *P*. Show the triangle in the positions both before and after rotation.

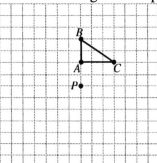

17. Construct a glide reflection of square *ABCD* shown below, using translation vector **v** and reflection line *l*. Show the square in the positions both before and after the glide reflection.

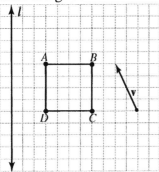

18. Use the figure below to answer the following questions.

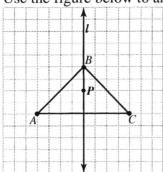

 a) Does triangle *ABC* have reflective symmetry about line *l*? Explain.
 b) Does triangle *ABC* have 180° rotational symmetry about point *P*? Explain.

19. What is a Klein bottle?

20. a) Sketch an object of genus 0.
  b) Sketch an object of genus 1.

21. State the fifth axiom of Euclidean geometry.

A Survey of Mathematics with Applications, 7e

Chapter 9 – Geometry   **Form 2**

In Exercises 1 – 4, use the following figure to describe the set of points.

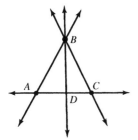

1. $\overset{\circ}{\overrightarrow{BA}} \cup \overrightarrow{BC}$

2. $\overleftrightarrow{AC} \cap \overrightarrow{CD}$

3. $\measuredangle ABD \cap \measuredangle CBD$

4. $\overline{AB} \cup \overline{BD} \cup \overline{DA}$

5. $m \measuredangle Q = 62.1°$. Determine the measure of the complement of $\measuredangle Q$.

6. $m \measuredangle Q = 102.5°$. Determine the measure of the supplement of $\measuredangle Q$.

7.   Determine the measure of $\measuredangle x$ in the following figure.

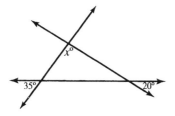

8.   Determine the sum of the measures of the interior angles of a Hexagon.

9.   Trapezoids $ABCD$ and $A'B'C'D'$ are similar figures. Determine the length of side $\overline{A'B'}$.

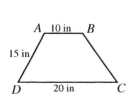 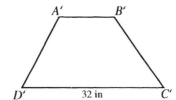

10. Right triangle *ABC* has one leg of length 9 ft and a hypotenuse of 15 ft.
    a) Determine the length of the other leg.
    b) Determine the perimeter of the triangle.
    c) Determine the area of the triangle.

11. Determine the volume of a cone having a base diameter of 9 cm and a height of 10 cm.

12. How much cubic inches of water will a soup can hold if it has a diameter of 6 in and a height of 8 in?

13. Determine the volume of a rectangular prism with a length of 6 m, a width of 2 m, and a height of 7 m.

14. Construct a reflection of triangle *ABC* shown below, about line *l*. Show the triangle in the positions both before and after reflection.

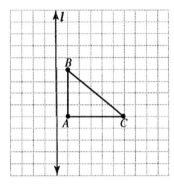

15. Construct a translation of the rectangle *ABCD* shown below using translation vector **v**. Show the rectangle in positions both before and after the translation.

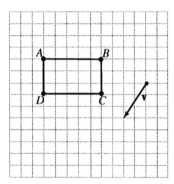

16. Construct a 270° rotation of triangle *ABC* shown below, about rotation point *P*. Show the triangle in the positions both before and after rotation.

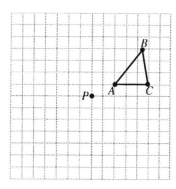

17. Construct a glide reflection of trapezoid *ABCD* shown below, using translation vector **v** and reflection line *l*. Show the trapezoid in the positions both before and after the glide reflection.

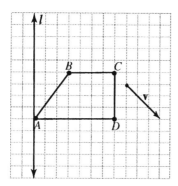

18. Use the figure below to answer the following questions.

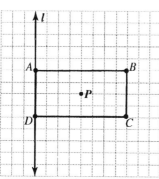

    a) Does rectangle *ABCD* have reflective symmetry about line *l*? Explain.
    b) Does rectangle *ABCD* have 180° rotational symmetry about point *P*? Explain.

19. What is a Jordan curve?

20. a) Sketch an object of genus 2.
   b) Sketch an object of genus 3 or more.

21. State the fifth axiom of Elliptical geometry.

A Survey of Mathematics with Applications, 7e

Chapter 9 – Geometry

Form 3

In Exercises 1 – 4, use the following figure to describe the set of points.

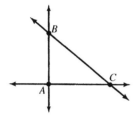

1. $\overrightarrow{AC} \cap \overrightarrow{AB}$

2. $\overset{\circ}{\overrightarrow{BC}} \cup \overrightarrow{BA}$

3. $\sphericalangle ABC \cap \sphericalangle BCA$

4. $\overset{\circ}{\overrightarrow{AC}} \cap \overset{\circ}{\overrightarrow{AB}}$

5. $m \sphericalangle D = 8.4°$. Determine the measure of the complement of $\sphericalangle D$.

6. $m \sphericalangle D = 99.9°$. Determine the measure of the supplement of $\sphericalangle D$.

7. Determine the measure of $\sphericalangle x$ in the following figure.

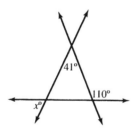

8. Determine the sum of the measures of the interior angles of a Heptagon.

9. Triangles ABC and $A'B'C'$ are similar figures. Determine the length of side $\overline{A'C'}$.

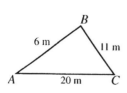
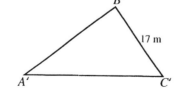

10. Right triangle *ABC* has one leg of length 6 ft and a hypotenuse of 10 ft.
   a) Determine the length of the other leg.
   b) Determine the perimeter of the triangle.
   c) Determine the area of the triangle.

11. Determine the volume of a cube having a side length of 11 ft.

12. How many cubic meters of concrete are needed to fill in a rectangular portion of sidewalk that is 3 m long, 2 m wide, and 0.2 m deep?

13. Determine the volume of a pyramid that has square base. The length of one side of the base is 6 in and the height of the pyramid is 11 in.

14. Construct a reflection of square *ABCD* shown below, about line *l*. Show the square in the positions both before and after reflection.

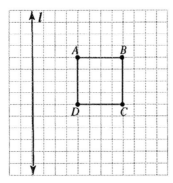

15. Construct a translation of the trapezoid *ABCD* shown below using translation vector **v**. Show the trapezoid in positions both before and after the translation.

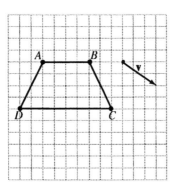

16. Construct a 180° rotation of square *ABCD* shown below, about rotation point *P*. Show the triangle in the positions both before and after rotation.

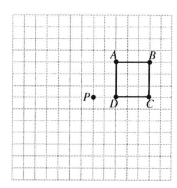

17. Construct a glide reflection of triangle *ABC* shown below, using translation vector **v** and reflection line *l*. Show the triangle in the positions both before and after the glide reflection.

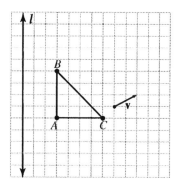

18. Use the figure below to answer the following questions.

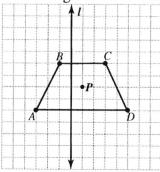

       a) Does trapezoid *ABCD* have reflective symmetry about line *l*? Explain.
       b) Does triangle *ABC* have 360° rotational symmetry about point *P*? Explain.

19. Explain how to make a Möbius strip?

20. a) Sketch an object of genus 0.
   b) Sketch an object of genus 2.

21. State the fifth axiom of Hyperbolic geometry.

A Survey of Mathematics with Applications, 7e

Chapter 10 – Mathematical Systems   Form 1

1. What is a binary operation?

2. Is it possible that a mathematical system is a group but not a commutative group? Explain.

3. Is the set of positive integers a group under the operation of subtraction? Explain your answer completely.

4. Develop a clock 4 arithmetic addition table.

5. Is clock 4 arithmetic a commutative group under the operation of addition? Assume that the associative property holds. Explain your answer completely.

In Exercises 6 and 7, determine the answer in clock 4 arithmetic.

6. $2 + 3 + 3$

7. $1 - 4$

8. Consider the mathematical system below.

✦	T	R	A	K
T	A	K	T	R
R	K	T	R	A
A	T	R	A	K
K	R	A	K	T

a) What is the binary operation?
b) Is this system closed? Explain.
c) Is there an identity element for this system under the given operation? If so, name it.
d) What is the inverse of element K?
e) What is $(T \divideontimes K) \divideontimes R$?

In Exercises 9 and 10, determine whether the mathematical system is a commutative group. Explain your answer completely.

9.

↗	J	Q	K
J	K	J	Q
Q	J	Q	K
K	Q	K	J

10.

◆	3	6	9
3	6	3	9
6	3	9	6
9	9	6	3

11. Determine whether the following mathematical system is a commutative group. Assume that the associative property holds. Explain your answer completely.

❖	!	?	<	>
!	<	>	!	?
?	>	!	?	<
<	!	?	<	>
>	?	<	>	!

In Exercises 12 and 13, determine the modulo class to which the number belongs for the indicated modulo system.

12. 81, modulo 10

13. 72, modulo 5

In Exercises 14 – 18, find all replacements for the question mark, less than the modulus, that make the statement true.

14. $? - 2 \equiv 2 \ (\text{mod } 3)$

15. $5 - ? \equiv 6 \ (\text{mod } 7)$

16. $5 + 10 \equiv ? \ (\text{mod } 12)$

17. $3 \cdot 5 \equiv ? \ (\text{mod } 8)$

18. $4 \cdot ? \equiv 2 \ (\text{mod } 5)$

19. What number is 125 congruent to in modulo 11?

20. a) Construct a modulo 4 multiplication table.
   b) Is this mathematical system a commutative group? Explain your answer completely.

A Survey of Mathematics with Applications, 7e

1. What is a counterexample?

2. Is it possible that a mathematical system is a commutative group, but not a group? Explain.

3. Is the set of integers closed under the operation of multiplication? Explain your answer completely.

4. Develop a clock 6 arithmetic addition table.

5. Is clock 6 arithmetic a commutative group under the operation of addition? Assume that the associative property holds. Explain your answer completely.

In Exercises 6 and 7, determine the answer in clock 6 arithmetic.

6. $4 + 10 + 1$

7. $2 - 8$

8. Consider the mathematical system below.

✿	C	L	A	Y
C	C	L	A	Y
L	L	A	Y	C
A	A	Y	C	L
Y	Y	C	L	A

   a) What is the binary operation?
   b) Is this system closed? Explain.
   c) Is there an identity element for this system under the given operation? If so, name it.
   d) What is the inverse of element A?
   e) What is L ✿ $(Y$ ✿ $Y)$?

In Exercises 9 and 10, determine whether the mathematical system is a commutative group. Explain your answer completely.

9.

♪	2	4	8
2	4	2	8
4	2	4	8
8	8	4	2

10.

✿	T	L	H
T	L	H	T
L	H	T	L
H	T	L	H

11. Determine whether the following mathematical system is a commutative group. Assume that the associative property holds. Explain your answer completely.

⤴	@	#	*	$
@	#	*	$	@
#	*	$	@	#
*	$	@	#	*
$	@	#	*	$

In Exercises 12 and 13, determine the modulo class to which the number belongs for the indicated modulo system.

12. 65, modulo 4

13. 57, modulo 5

In Exercises 14 – 18, find all replacements for the question mark, less than the modulus, that make the statement true.

14. $2 - ? \equiv 3 \pmod 4$

15. $? - 7 \equiv 4 \pmod 6$

16. $11 + 6 \equiv ? \pmod{10}$

17. $6 \cdot ? \equiv 2 \pmod 3$

18. $12 \cdot 2 \equiv ? \pmod 7$

19. What number is 132 congruent to in modulo 5?

20. a) Construct a modulo 6 multiplication table.
 b) Is this mathematical system a commutative group? Explain your answer completely.

Chapter 10 – Mathematical Systems   Form 3

1. What is another name for a commutative group?

2. Explain the closure property, and give an example of the property.

3. Is the set of natural numbers closed under the operation of division? Explain your answer completely.

4. Develop a clock 3 arithmetic addition table.

5. Is clock 3 arithmetic a commutative group under the operation of addition? Assume that the associative property holds. Explain your answer completely.

In Exercises 6 and 7, determine the answer in clock 3 arithmetic.

6. $4 + 3 + 6$

7. $3 - 5$

8. Consider the mathematical system below.

✱	J	S	H	B
J	H	B	J	S
S	B	J	S	H
H	J	S	H	B
B	S	H	B	J

  a) What is the binary operation?
  b) Is this system closed? Explain.
  c) Is there an identity element for this system under the given operation? If so, name it.
  d) What is the inverse of element *S*?
  e) What is $J ✱ (S ✱ H)$?

In Exercises 9 and 10, determine whether the mathematical system is a commutative group. Explain your answer completely.

9.

♉	?	∶	!
?	!	?	∶
∶	?	∶	!
!	∶	!	?

10.

♥	A	B	C
A	B	A	C
B	A	B	C
C	C	B	A

11. Determine whether the following mathematical system is a commutative group. Assume that the associative property holds. Explain your answer completely.

≋	2	4	6	8
2	4	6	8	2
4	6	8	2	4
6	8	2	4	6
8	2	4	6	8

In Exercises 12 and 13, determine the modulo class to which the number belongs for the indicated modulo system.

12. 25, modulo 3

13. 68, modulo 7

In Exercises 14 – 18, find all replacements for the question mark, less than the modulus, that make the statement true.

14. $18 + 5 \equiv ? \ (\text{mod } 11)$

15. $? - 6 \equiv 3 \ (\text{mod } 8)$

16. $4 - ? \equiv 2 \ (\text{mod } 4)$

17. $5 \cdot 2 \equiv ? \ (\text{mod } 7)$

18. $3 \cdot ? \equiv 3 \ (\text{mod } 5)$

19. What number is 210 congruent to in modulo 13?

20. a) Construct a modulo 3 multiplication table.
   b) Is this mathematical system a commutative group? Explain your answer completely.

A Survey of Mathematics with Applications, 7e

Chapter 11 – Consumer Math   **Form 1**

In Exercise 1 and 2, find the missing quantity by using the simple interest formula.

1. $i = ?$, $p = \$6000$, $r = 2\%$ per year, $t = 4$ months

2. $i = \$120$, $p = \$800$, $r = ?$ per year, $t = 3$ months

In Exercises 3 and 4, Jacob Rosenberg borrowed $6400 from a bank for 15 months. The rate of simple interest charged is 4.5%.

3. How much interest did he pay for the use of the money?

4. What is the amount he repaid the bank on the due date of the loan?

In Exercises 5 and 6, Lee Lai Hoang received an $8500 loan with interest at 9% for 126 days on May 4. Lee Lai made a payment of $4000 on July 2.

5. How much did she owe the bank on the date of maturity?

6. What total amount of interest did she pay on the loan?

In Exercises 7 and 8, compute the amount and the compound interest.

	Principal	Time	Rate	Compounded
7.	$6000	4 years	8.5%	Semiannually
8.	$3500	2 years	4%	Quarterly

In Exercises 9 – 11, a new DVD camcorder sells for $1250. To finance the camcorder through a bank, the bank requires a down payment of 10% and monthly payments of $96.50 for 12 months.

9. How much money will the purchaser have to borrow from the bank?

10. What finance charge will the purchaser have to pay the bank?

11. What is the APR?

12. Sean Taylor purchased a new couch and love seat for his living room for $4500. He made a down payment of $1550 and financed the balance with a 24-month fixed-payment installment loan. Instead of making the eleventh payment of $144.17, Sean decides to pay off the loan.
    a) How much interest will Sean save (use the rule of 78s)?
    b) What is the total amount due to pay off the loan?

13. Victor Gil borrowed $10,500. To repay the loan, he was scheduled to make 48 monthly installment payments of $243.23 each. Instead of making his 36th payment, Victor decides to pay off the loan.

    a) Determine the APR of the installment loan.
    b) How much interest will Victor save (use the actuarial method)?
    c) What is the total amount due to pay off the loan?

14. Fantasia Johnson's credit card statement shows a balance due of $567.50 on August 24, the billing date. For the period ending on September 24, she had the following transactions.

August 28	Charge: Restaurant	$75.26
September 5	Payment	380.50
September 8	Charge: Shoes	59.60
September 18	Charge: Software	105.89
September 21	Charge: Body Jewelry	110.57

    a) Find the finance charge on September 24 by using the unpaid balance method. Assume that the interest rate is 1.5% per month.
    b) Find the new account balance on September 24 using the finance charge found in part (a).
    c) Find the average daily balance for the period.
    d) Find the finance charge on September 24 by using the average daily balance method. Assume that the interest rate is 1.5% per month.
    e) Find the new account balance on September 24 using the finance charge found in part (d).

In Exercises 15 – 21, the Liebermans have found their dream house. It costs $250,000. The taxes on the house would be $4000 per year, and insurance would cost $950 per year. The Liebermans have applied for a conventional loan at the bank. The bank is requiring a 30% down payment, and the interest rate is 9%. The Liebermans' annual income is $98,500. They have more than 10 monthly payments remaining on each of the following: $440 for a car and $275 for a plasma screen television.

15. What is the required down payment?

16. Determine their adjusted monthly income.

17. What is the maximum monthly payment the bank's loan officer believes the Liebermans can afford?

18. Determine the monthly payments of principal and interest for a 25-year loan.

19. Determine their total monthly payments, including insurance and taxes.

20. Does the bank's loan officer believe that the Liebermans meet the requirements for the mortgage?

21. a) Find the total cost of the house (excluding insurance and taxes) after 25 years.
b) How much of the total cost is interest?

A Survey of Mathematics with Applications, 7e

Chapter 11 – Consumer Math   Form 2

In Exercise 1 and 2, find the missing quantity by using the simple interest formula.

1. $i = ?$, $p = \$785$, $r = 6\%$ per year, $t = 8$ months

2. $i = \$38.25$, $p = \$1700$, $r = 4.5\%$ per year, $t = ?$

In Exercises 3 and 4, Sarah Jamberger borrowed $2500 from a bank for 18 months. The rate of simple interest charged is 5.5%.

3. How much interest did she pay for the use of the money?

4. What is the amount she repaid the bank on the due date of the loan?

In Exercises 5 and 6, John Kelly received an $6700 loan with interest at 11% for 200 days on March 27. John made a payment of $3200 on August 12.

5. How much did he owe the bank on the date of maturity?

6. What total amount of interest did he pay on the loan?

In Exercises 7 and 8, compute the amount and the compound interest.

	Principal	Time	Rate	Compounded
7.	$7000	3 years	3.5%	Monthly
8.	$1250	6 years	8%	Annually

In Exercises 9 – 11, a new camera cell phone sells for $325. To finance the cell phone through a bank, the bank requires a down payment of 5% and monthly payments of $13.50 for 24 months.

9. How much money will the purchaser have to borrow from the bank?

10. What finance charge will the purchaser have to pay the bank?

11. What is the APR?

12. Jodie Fry purchased a new computer that sells for $3200. She made a down payment of $750 and financed the balance with a 12-month fixed-payment installment loan. Instead of making the eighth payment of $287.50, she decides to pay off the loan.
    a) How much interest will Jodie save (use the rule of 78s)?
    b) What is the total amount due to pay off the loan?

13. Oded Fehr borrowed $11,000. To repay the loan, he was scheduled to make 36 monthly installment payments of $340 each. Instead of making his 24th payment, Oded decides to pay off the loan.

    a) Determine the APR of the installment loan.
    b) How much interest will Oded save (use the actuarial method)?
    c) What is the total amount due to pay off the loan?

14. Yolanda Murphy's credit card statement shows a balance due of $286.75 on September 17, the billing date. For the period ending on October 17, she had the following transactions.

September 20	Charge: Television	$825
September 29	Charge: Crib	650
October 1	Payment	600
October 10	Charge: Baby Monitor	50
October 15	Charge: Groceries	112

    a) Find the finance charge on October 17 by using the unpaid balance method. Assume that the interest rate is 1.2% per month.
    b) Find the new account balance on October 17 using the finance charge found in part (a).
    c) Find the average daily balance for the period.
    d) Find the finance charge on October 17 by using the average daily balance method. Assume that the interest rate is 1.2% per month.
    e) Find the new account balance on October 17 using the finance charge found in part (d).

In Exercises 15 – 21, Carlos Singer inherited money from a relative and decided to buy his first house. It is a townhouse which costs $175,000. The taxes on the house would be $2000 per year, and insurance would cost $750 per year. Carlos has applied for a conventional loan at the bank. The bank is requiring a 15% down payment, and the interest rate is 8.5%. Carlos' annual income is $88,750. He has more than 10 monthly payments remaining on each of the following: $150 for a computer and $285 for a car.

15. What is the required down payment?

16. Determine their adjusted monthly income.

17. What is the maximum monthly payment the bank's loan officer believes Carlos can afford?

18. Determine the monthly payments of principal and interest for a 20-year loan.

19. Determine their total monthly payments, including insurance and taxes.

20. Does the bank's loan officer believe that Carlos meets the requirements for the mortgage?

21. a) Find the total cost of the house (excluding insurance and taxes) after 20 years.
b) How much of the total cost is interest?

A Survey of Mathematics with Applications, 7e

Chapter 11 – Consumer Math   Form 3

In Exercise 1 and 2, find the missing quantity by using the simple interest formula.

1. $i = ?$, $p = \$485$, $r = 4.5\%$ per year, $t = 8$ months

2. $i = \$178.50$, $p = ?$, $r = 6\%$ per year, $t = 7$ months

In Exercises 3 and 4, Blossom Meir borrowed $5750 from a bank for 17 months. The rate of simple interest charged is 3.6%.

3. How much interest did she pay for the use of the money?

4. What is the amount she repaid the bank on the due date of the loan?

In Exercises 5 and 6, Richie Ashburn received an $7500 loan with interest at 10.5% for 180 days on December 1. Richie made a payment of $3000 on February 27.

5. How much did he owe the bank on the date of maturity?

6. What total amount of interest did he pay on the loan?

In Exercises 7 and 8, compute the amount and the compound interest.

	Principal	Time	Rate	Compounded
7.	$5500	2 years	1.7%	Quarterly
8.	$9600	6 years	8%	Semiannually

In Exercises 9 – 11, a new plasma screen television sells for $3700. To finance the television through a bank, the bank requires a down payment of 20% and monthly payments of $175 for 18 months.

9. How much money will the purchaser have to borrow from the bank?

10. What finance charge will the purchaser have to pay the bank?

11. What is the APR?

12. Sofia Micelli purchased ceramic tile for her new home for $4800. She made a down payment of $1800 and financed the balance with a 30-month fixed-payment installment loan. Instead of making the eighteenth payment of $135, Sofia decides to pay off the loan.
    a) How much interest will Sofia save (use the rule of 78s)?
    b) What is the total amount due to pay off the loan?

13. Kim Wong borrowed $7000. To repay the loan, she was scheduled to make 36 monthly installment payments of $217.36 each. Instead of making his 18th payment, Kim decides to pay off the loan.

   a) Determine the APR of the installment loan.
   b) How much interest will Kim save (use the actuarial method)?
   c) What is the total amount due to pay off the loan?

14. Henry Martel's credit card statement shows a balance due of $875 on February 13, the billing date. For the period ending on March 13, he had the following transactions. Assume it is not a leap year.

February 18	Charge:	Car Tires	$440.25
February 26	Charge:	Ties	85.10
March 1	Payment		600.50
March 5	Charge:	Restaurant	125.82
March 10	Charge:	Cologne	56.73

   a) Find the finance charge on March 13 by using the unpaid balance method. Assume that the interest rate is 2% per month.
   b) Find the new account balance on March 13 using the finance charge found in part (a).
   c) Find the average daily balance for the period.
   d) Find the finance charge on March 13 by using the average daily balance method. Assume that the interest rate is 2% per month.
   e) Find the new account balance on March 13 using the finance charge found in part (d).

In Exercises 15 – 21, the Nemeths decided to buy a bigger house. It costs $310,000. The taxes on the house would be $4700 per year, and insurance would cost $1100 per year. The Nemeths have applied for a conventional loan at the bank. The bank is requiring a 35% down payment, and the interest rate is 7.5%. The Nemeths' annual income is $110,000. They have more than 10 monthly payments remaining on each of the following: $350 for a college education loan and $375 for a car.

15. What is the required down payment?

16. Determine their adjusted monthly income.

17. What is the maximum monthly payment the bank's loan officer believes the Nemeths can afford?

18. Determine the monthly payments of principal and interest for a 30-year loan.

19. Determine their total monthly payments, including insurance and taxes.

20. Does the bank's loan officer believe that the Nemeths meet the requirements for the mortgage?

21. a) Find the total cost of the house (excluding insurance and taxes) after 30 years.
b) How much of the total cost is interest?

A Survey of Mathematics with Applications, 7e

Chapter 12 – Probability

1. Fifty people who visited the Humane Society adopted a pet. Twenty of them adopted a cat. Find the empirical probability that the next person who adopts a pet at the Humane Society adopts a cat.

In Exercise 2 – 5, each of the numbers 5 – 15 is written on a sheet of paper. The eleven sheets of paper are placed in a small wicker basket. If one sheet of paper is selected at random from the basket, find the probability that the number selected is

2. less than 10.

3. even.

4. odd or greater than 8.

5. even and greater than 8.

In Exercises 6 – 9, if two of the same eleven sheets of paper mentioned in the above exercises are selected from the basket, without replacement, find the probability that

6. both numbers are less than 10.

7. both numbers are odd.

8. the first number is even and the second number is greater than 10.

9. neither number is odd.

10. One card is selected at random from a deck of cards. Find the probability that the card selected is a spade or a queen.

In Exercises 11 and 12, one colored marble – purple, green, red, or yellow – is selected at random and a coin is tossed.

11. Use the counting principle to determine the number of sample points in the sample space.

12. Construct a tree diagram illustrating all the possible outcomes, and list the sample space.

In Exercises 13 – 15, by observing the sample space of the marbles and coin in Exercise 12, determine the probability of obtaining

13. the color red and a head.

14. the color purple or a tail.

15. a color other than yellow and a tail.

16. An email password is to consist of five letters followed by two digits. Find the number of passwords possible if the first letter cannot be X, Y, or Z, and repetition is permitted.

17. An elementary school class consists of 12 girls and 16 boys. If one student is selected at random, find the odds

      a) in favor of the student being a boy.
      b) against the student being a boy.

18. The odds in favor of Donovan making the football team are 7:2. Find the probability that Donovan does not make the team.

19. You get to roll a die one time. If you roll an odd number, you win $3. If you roll a 6, you win $10. If you roll any other number, you lose $4. Find the expectation for this game.

20. Lisa's School of Dance keeps a record of their students' ages and the dance classes they prefer taking. The results are shown below.

Age	Ballet	Jazz	Tap	Total
3 – 6	25	20	20	65
7 – 12	30	60	45	135
13 or older	45	50	30	125
Total	**100**	**130**	**95**	**315**

If one student is selected at random, find the probability that

      a) the student is 13 or older.
      b) the student prefers taking jazz.
      c) the student prefers taking tap, given the student is 3 – 6 years old.
      d) the student is 7 – 12 years old, given the student prefers taking ballet.

21. Al's Appliances employs ten salespersons. To motivate sales, the store is going to award prizes to the top three salespersons of the month. First prize is a stereo system, second prize is a personal MP3 player, and third prize is a alarm clock radio. In how many ways can these prizes be awarded?

In Exercises 22 and 23, a garment rack contains 12 blouses, of which 7 are beige and 5 are white. If you select 3 at random, without replacement, find the probability that:

22. None of the blouses is beige.

23. At least one blouse is beige.

24. Ten pencils and eight pens are in a box. Eleven writing utensils are to be selected at random, without replacement, from the box. If the pens and pencils are equally likely to be selected, determine the probability that seven pencils and four pens are selected.

25. The probability of a male student wearing a baseball cap to class is 0.2. Find the probability that exactly two out of eight male students will be wearing a baseball cap to class.

A Survey of Mathematics with Applications, 7e

Chapter 12 – Probability

Form 2

1. Out of 60 people that have gone car shopping at Ed Morse Chevrolet this Labor Day Weekend, 15 of them have purchased a car. Find the empirical probability that the next person to go car shopping at Ed Morse Chevrolet this Labor Day weekend will buy a car.

In Exercise 2 – 5, each of the numbers 0 – 9 is written on a different golf ball. The ten golf balls are put in a paper bag. If one golf ball is selected at random from the bag, find the probability that the number selected is

2. greater than 5.

3. odd.

4. odd or less than 8.

5. even and greater than 4.

In Exercises 6 – 9, if two of the same ten golf balls mentioned in the above exercises are selected from the bag, without replacement, find the probability that

6. both numbers are greater than 5.

7. both numbers are less than 4.

8. the first number is odd and the second number is greater than 2.

9. neither number is less than 5.

10. One card is selected at random from a deck of cards. Find the probability that the card selected is a black card or a picture card.

In Exercises 11 and 12, one coin is tossed and one die is rolled.

11. Use the counting principle to determine the number of sample points in the sample space.

12. Construct a tree diagram illustrating all the possible outcomes, and list the sample space.

In Exercises 13 – 15, by observing the sample space of the coin and die in Exercise 12, determine the probability of obtaining

13. a tail and an even number.

14. a head or an odd number.

15. a head and a number less than 3.

16. A license plate is to consist of two letters followed by four digits. Find the number of license plates possible if the first letter cannot be Z, and repetition is permitted.

17. If a zoo consists of 5 tigers, 3 lions, and 2 bears, find the odds
      a) in favor of seeing a bear.
      b) against seeing a tiger.

18. The odds in favor of thunderstorms today are 2:5. Find the probability that there will not be any thunderstorms today.

19. You get to select one card at random from a deck of cards. If you pick any picture card, you win $4. If you pick a number less than five in any suit, you win $8. If you pick any other card, you lose $2. Find the expectation for this game.

20. The number of cars and work vans that use the Florida Turnpike Sunpass Lanes in a day is recorded. The results are shown below.

Sunpass	Cars	Work Vans	Total
Uses	1450	2500	3950
Does Not Use	850	3200	4050
Total	2300	5700	8000

If one vehicle is selected at random, find the probability that
      a) the vehicle does not use the Sunpass Lanes.
      b) the vehicle is a car.
      c) the vehicle uses the Sunpass Lanes, given it is a car.
      d) the vehicle does not use the Sunpass Lanes, given it is a work van.

21. Three people out of a twelve person committee can hold an office. One person can be president, another person can be vice president, and another person can be secretary. In how many different ways can these offices be held?

In Exercises 22 and 23, a garden has a total of 15 flowers, of which 10 are in bloom. If you select 2 at random, without replacement, find the probability that:

22. None of the flowers are in bloom.

23. At least one flower is in bloom.

24. Six green balloons and ten red balloons are in a bin. Eight balloons are to be selected at random, without replacement, from the bin. If the green and red balloons are equally likely to be selected, determine that four green balloons and four red balloons are selected.

25. The probability of finding a dog in a dog park on a given day that is not wearing a collar is 0.15. Find the probability that exactly three out of five dogs will not be wearing a collar in the dog park.

Chapter 12 – Probability   Form 3

1. Five people out of every fifteeen people who visit the doctor will leave with a prescription for a medication. Find the empirical probability that the next person who visits the doctor will leave with a prescription for a medication.

In Exercise 2 – 5, each of the numbers 1 – 15 is written on a different balloon. The fifteen balloons are then blown up and put in a large plastic bag. If one balloon is selected at random from the bag, find the probability that the number selected is

2. greater than 3.

3. even.

4. even and greater than 7.

5. odd or less than 5.

In Exercises 6 – 9, if two of the same fifteen balloons mentioned in the above exercises are selected from the bag at random, without replacement, find the probability that

6. both numbers are even.

7. both numbers are greater than 4.

8. neither number is even.

9. the first number is less than 6 and the second number is odd.

10. One card is selected at random from a deck of cards. Find the probability that the card selected is a red card or a ten.

In Exercises 11 and 12, one tie – dotted, striped, or solid – is selected at random and a pair of socks – white or black – is selected at random.

11. Use the counting principle to determine the number of sample points in the sample space.

12. Construct a tree diagram illustrating all the possible outcomes, and list the sample space.

In Exercises 13 – 15, by observing the sample space of the ties and socks in Exercise 12, determine the probability of obtaining

13. a dotted tie and white socks.

14. a striped tie or black socks.

15. a solid tie and white socks.

16. A seven digit phone number is to be chosen. Find the amount of possible phone numbers if the first digit cannot be zero and repetition is permitted.

17. A standard rack of 15 billiard balls contains 7 striped balls and 8 solid balls. If one billiard ball is selected at random, find the odds
   a) in favor of the ball being solid.
   b) against the ball being solid.

18. The odds against the horse, Golda's Revenge, winning the first face is 4:5. Find the probability that Golda's Revenge wins the first race.

19. You get to spin a wheel one time. Half of the wheel is green, a quarter of the wheel is blue, and a quarter of the wheel is red. Assume that you cannot land on a line in between colors. If the spinner lands on green, you win $1. If the spinner lands on blue, you win $5. If the spinner lands on red, you lose $2. Find the expectation for this game.

20. Amy surveyed the students in her class as to whether they preferred the color green or the color yellow. The results are shown below.

Student's Gender	Girl	Boy	Total
Prefers Green	12	10	22
Prefers Yellow	11	7	18
Total	23	17	40

If one student is selected at random, find the probability that
   a) the student is a girl.
   b) the student prefers green.
   c) the student prefers yellow, given the student is a boy.
   d) the student is a boy, given the student prefers green.

21. A royal flush consists of an ace, king, queen, jack, and 10 all in the same suit. If 5 cards are dealt at random from a pile of cards consisting of the suit of hearts (13 cards), find the probability of getting a royal flush in hearts.

In Exercises 22 and 23, a box consists of 12 light bulbs of which 3 are defective. If you select 2 at random, without replacement, find the probability that:

22. None of the light bulbs are defective.

23. At least one of the light bulbs are defective.

24. Ten pairs of jeans and 12 t-shirts are in a trunk. Ten pieces of clothing are to be selected at random, without replacement, from the trunk. If the jeans and t-shirts are equally likely to be selected, determine the probability that 4 pairs of jeans and 6 t-shirts are selected.

25. The probability of passing a certain English class is 0.7. Find the probability that exactly five out of fifteen students will pass this English class.

Chapter 13 – Statistics   Form 1

In Exercises 1 – 6, for the set of data 6, 10, 15, 20, 20, determine the:

1. mean

2. median

3. mode

4. midrange

5. range

6. standard deviation

In Exercises 7 – 9, use the following set of data to construct

12	18	31	42
15	25	35	42
15	25	37	43
15	30	38	48
18	30	39	48

7. a frequency distribution; let the first class be 10 – 15.

8. a histogram of the frequency distribution.

9. a frequency polygon of the frequency distribution.

In Exercises 10 – 16, use the following data on bowlers' scores at an amateur tournament.

Mean	166	First quartile	150
Median	175	Third quartile	190
Mode	150	88th percentile	205
Standard deviation	33		

10. What is the most common bowling score?

11. What score do half the bowlers exceed?

12. About what percent of the scores exceed 150?

13. About what percent of the scores are below 205?

14. If there were 50 games played in the tournament, what would be the total of all scores for all the games?

15. What score represents two standard deviations above the mean?

16. What score represents one standard deviation below the mean?

In Exercises 17 – 20, use the following information: The selling price for houses in Sunrise, Florida is normally distributed with a mean of $250,000 and a standard deviation of $30,000.

17. What percent of the houses sold were between $255,000 and $260,000?

18. What percent of the houses sold for greater than $200,000?

19. What percent of the houses sold for greater than $270,000?

20. If a random sample of 550 houses were selected, how many would have sold between $210,000 and $240,000?

21. The following chart shows the pounds of coffee brewed per day in different sized coffee shops.

Size (in square yards)	Pounds of Coffee Brewed
30	5
44	9
57	18
66	23
106	31

a) Construct a scatter diagram placing the size on the horizontal axis.
b) Use the scatter diagram in part (a) to determine whether you believe that a correlation exists between the size of a coffee shop and the pounds of coffee brewed daily. Explain.
c) Determine the correlation coefficient between the size of a coffee shop and the pounds of coffee brewed daily.
d) Determine whether a correlation exists at $\alpha = 0.05$
e) Assuming that this trend continues, determine the equation of the line of best fit between the size of a coffee shop and the pounds of coffee brewed daily.
f) Use the equation in part (e) to predict the pounds of coffee brewed daily in a coffee shop that is 95 square yards.

A Survey of Mathematics with Applications, 7e

Chapter 13 – Statistics   Form 2

In Exercises 1 – 6, for the set of data 2, 6, 6, 18, 25, 28, determine the:

1. mean

2. median

3. mode

4. midrange

5. range

6. standard deviation

In Exercises 7 – 9, use the following set of data to construct

25	32	36	42
26	33	38	45
28	34	40	45
28	35	41	47
30	35	41	48

7. a frequency distribution; let the first class be 25 – 30.

8. a histogram of the frequency distribution.

9. a frequency polygon of the frequency distribution.

In Exercises 10 – 16, use the following data on the number of houses built weekly by the Pyramid Construction Company.

Mean	500	First quartile	420
Median	470	Third quartile	510
Mode	485	53rd percentile	468
Standard deviation	60		

10. What is the most common number of houses built weekly?

11. How many houses built exceed the halfway point?

12. About what percent of houses built exceed 420?

13. About what percent of houses built is less than 468?

14. In 6 weeks, what would be the total number of houses built?

15. What number of houses built represents two standard deviations below the mean?

16. What number of houses built represents one standard deviation above the mean?

In Exercises 17 – 20, use the following information: The mileage at which a car is due for an oil change is normally distributed with a mean of 3000 miles and a standard deviation of 200 miles.

17. What percent of cars have an oil change at a mileage between 2800 and 3100?

18. What percent of cars have an oil change at a mileage less than 3300?

19. What percent of cars have an oil change at a mileage less than 2700?

20. If a random sample of 600 cars were selected, how many of them would have an oil change at a mileage between 2600 and 2900?

21. The following chart shows the number of concert tickets purchased at a small venue and the units of soda sold.

Tickets Purchased	Soda Sold
100	15
150	20
200	37
250	48
300	115

a) Construct a scatter diagram placing the number of concert tickets on the horizontal axis.

b) Use the scatter diagram in part (a) to determine whether you believe that a correlation exists between the number of concert tickets purchased and the units of soda sold. Explain.

c) Determine the correlation coefficient between the number of concert tickets purchased and the units of soda sold.

d) Determine whether a correlation exists at $\alpha = 0.01$

e) Assuming that this trend continues, determine the equation of the line of best fit between the number of concert tickets purchased and the units of soda sold.

f) Use the equation in part (e) to predict the units of soda sold at a concert where 175 tickets were purchased.

A Survey of Mathematics with Applications, 7e

In Exercises 1 – 6, for the set of data 7, 7, 10, 12 14, determine the:

1. mean

2. median

3. mode

4. midrange

5. range

6. standard deviation

In Exercises 7 – 9, use the following set of data to construct

50	67	81	90
56	69	81	95
56	72	85	98
58	72	87	98
60	80	88	99

7. a frequency distribution; let the first class be 50 – 59.

8. a histogram of the frequency distribution.

9. a frequency polygon of the frequency distribution.

In Exercises 10 – 16, use the following data on students' SAT scores.

Mean	980	First quartile	750
Median	1000	Third quartile	1250
Mode	850	90th percentile	1400
Standard deviation	50		

10. What is the most common SAT score?

11. What score do half the students exceed?

12. About what percent of the scores exceed 750?

13. About what percent of the scores are below 1400?

14. If 200 students take the SAT on a particular day, what would be the total of all the scores for that day?

15. What score represents 1.5 standard deviations above the mean?

16. What score represents one standard deviation below the mean?

In Exercises 17 – 20, use the following information: The weight of men between the ages of 18 and 74 are normally distributed with a mean of 170 pounds and a standard deviation of 20 pounds.

17. What percent of men weigh between 190 and 215 pounds?

18. What percent of men weight more than 155 pounds?

19. What percent of men weigh less than 200 pounds?

20. If a random sample of 600 men were selected, how many men would weigh between 155 and 180 pounds?

21. The following chart shows the number of years of college attended and the amount of annual income.

Years of College Attendance	Annual Income (In Thousands)
0	$25
2	$40
4	$60
7	$85
11	$100

a) Construct a scatter diagram placing the years of college attendance on the horizontal axis.
b) Use the scatter diagram in part (a) to determine whether you believe that a correlation exists between the years of college attendance and amount of annual income. Explain.
c) Determine the correlation coefficient between the years of college attendance and the amount of annual income.
d) Determine whether a correlation exists at $\alpha = 0.05$
e) Assuming that this trend continues, determine the equation of the line of best fit between the years of college attendance and the amount of annual income.
f) Use the equation in part (e) to predict the amount of annual income for nine years of college attendance.

A Survey of Mathematics with Applications, 7e

Chapter 14 – Graph Theory   Form 1

1. Create a graph with six vertices and a bridge.

2. Represent the map below as a graph where each vertex represents a state and each edge represents a common border between the states.

3. Draw a connected graph.

4. In the following graph, determine an Euler path.

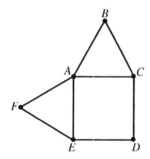

5. Is it possible for a person to walk through each doorway in the house, whose floor plan is shown below, without using any of the doorways twice? If so, indicate in which room the person may start and in which room the person may finish.

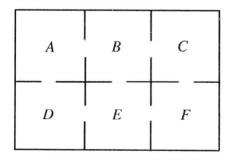

6. Use Fluery's algorithm to determine an Euler circuit in the following graph.

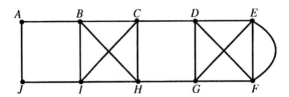

7. Determine a Hamilton Circuit in the following graph.

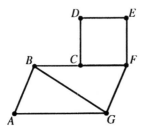

8. Enrique Rodriguez-Paz lives in Tamarac, Florida and has job interviews in the following other Florida cities: Orlando, Tampa, Miami, Naples, Fort Pierce, and Stuart. How many different ways can Enrique visit each city and return to his home in Tamarac?

9. Jennie Tartaglia lives in New York City and has job interviews in Atlanta, Georgia; San Diego, California; and Reno, Nevada. The one-way flights between these four cities is as follows: New York City to Atlanta is $175, New York City to San Diego is $400, New York City to Reno is $350, Atlanta to San Diego is $300, Atlanta to Reno is $200, and San Diego to Reno is $75.

   a) Represent this traveling salesman problem with a complete, weighted graph showing prices of flights on the appropriate edges.
   b) Use the Brute Force method to determine the least expensive route for Jennie to visit each city once and return home to New York City. What is the cost when using this route?
   c) Use the Nearest Neighbor method to approximate the optimal route for Jennie to visit each city once and return home to New York City. What is the cost when using this route?

10. Determine a spanning tree for the graph shown below.

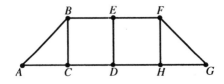

11. Determine the minimum-cost spanning tree for the following weighted graph.

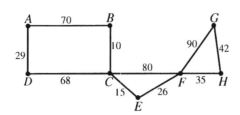

12. Tara Lee is setting up a computer network system for her new offices. Her current system has the computers in place as shown in the figure below. The numbers shown are in inches.

a) Determine the minimum-cost spanning tree that reaches each computer.

b) If the new cable used to connect the computers cost $0.75 an inch, find the cost of setting up the system determined in part (a).

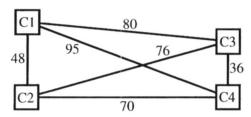

A Survey of Mathematics with Applications, 7e

Chapter 14 – Graph Theory   Form 2

1. Create a graph with 5 vertices, 2 bridges and a loop.

2. Represent the map below as a graph where each vertex represents a county and each edge represents a common border between the counties.

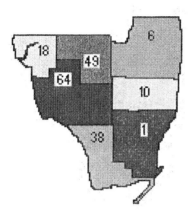

3. Draw a disconnected graph.

4. In the following graph, determine an Euler path.

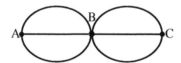

5. Is it possible for a person to walk through each doorway in the house, whose floor plan is shown below, without using any of the doorways twice? If so, indicate in which room the person may start and in which room the person may finish.

A	B	C
D	E	F

6. Use Fluery's algorithm to determine an Euler circuit in the following graph.

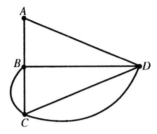

7. Determine a Hamilton Circuit in the following graph.

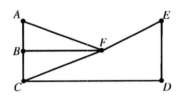

8. Oda Mae Brown lives in Brooklyn, New York, and will be visiting family in the following other New York boroughs: The Bronx, Manhattan, Queens, and Staten Island. How many different ways can Oda Mae visit each borough and return to her home in Brooklyn?

9. Starting from home, Julio Frías has several errands to run. He must stop at work to pick up some files, go to the drugstore to buy shaving products, and stop at the florist to get flowers for his girlfriend. Julio estimates the distances among these locations as follows: Home to work is 10 miles, home to the drugstore is 2 miles, home to the florist is 5 miles, work to the drugstore is 15 miles, work to the florist is 8 miles, and the drugstore to the florist is 11 miles.

   a) Represent this traveling salesman problem with a complete, weighted graph showing the distances between locations on the appropriate edges.
   b) Use the Brute Force method to determine the shortest route for Julio to run his errands and return home.
   c) Use the Nearest Neighbor method to approximate the optimal route for Julio to run each errand and return home. What is the distance traveled when using this route?

10. Determine a spanning tree for the graph shown below.

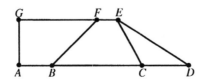

11. Determine the minimum-cost spanning tree for the following weighted graph.

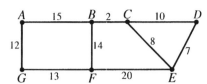

12. Sean and Shannon Murphy are installing an irrigation system to water their fruit and vegetable gardens. The distances in feet are shown below. is setting up a computer network system for her new offices. Their current yard has the gardens in place as shown in the figure below.

a) Determine the minimum-cost spanning tree that reaches each garden.

b) If the new irrigation pipes used to connect the gardens cost $5.55 per foot, find the cost of setting up the system determined in part (a).

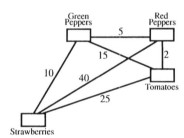

A Survey of Mathematics with Applications, 7e

Chapter 14 – Graph Theory   Form 3

1. Create a graph with five vertices and one loop.

2. Represent the map below as a graph where each vertex represents a state and each edge represents a common border between the states.

3. Draw a connected graph.

4. In the following graph, determine an Euler path.

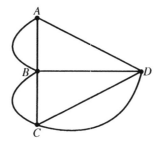

5. Is it possible for a person to walk through each doorway in the house, whose floor plan is shown below, without using any of the doorways twice? If so, indicate in which room the person may start and in which room the person may finish.

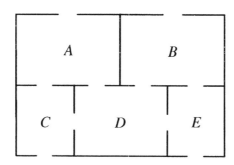

6. Use Fluery's algorithm to determine an Euler circuit in the following graph.

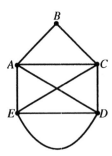

7. Determine a Hamilton Circuit in the following graph.

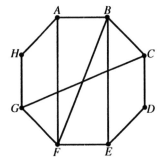

8. Rocco DiSpirito lives in New York City, New York, and is traveling to the following cities to promote his new book: Boston, Philadelphia, Newark, and Richmond. How many different ways can Rocco visit each city and return to his home in New York City?

9. Colleen Skoglund lives in Seattle, Washington. She is an American Idol fanatic and wants to visit the hometowns of some of the former contestants. Colleen will visit Raleigh, North Carolina, Birmingham, Alabama, and Dallas, Texas. The one-way flights between these four cities are as follows: Seattle to Raleigh is $420, Seattle to Birmingham is $360, Seattle to Dallas is $200, Raleigh to Birmingham is $85, Raleigh to Dallas is $125, and Birmingham to Dallas is $99.

   a) Represent this traveling salesman problem with a complete, weighted graph showing prices of flights on the appropriate edges.
   b) Use the Brute Force method to determine the least expensive route for Colleen to visit each city once and return home to Seattle. What is the cost when using this route?
   c) Use the Nearest Neighbor method to approximate the optimal route for Colleen to visit each city once and return home to Seattle. What is the cost when using this route?

10. Determine a spanning tree for the graph shown below.

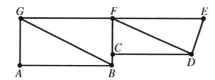

11. Determine the minimum-cost spanning tree for the following weighted graph.

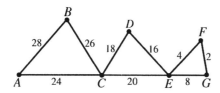

12. The Greensteins are getting a new television satellite system and must rewire the four televisions in their home. The televisions are already in place as shown below. The distances shown are in feet.

a) Determine the minimum-cost spanning tree that reaches each television.
b) If the new wiring used to connect the televisions cost $2.75 per foot, find the cost of setting up the system determined in part (a).

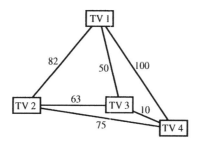

In Exercise 1 – 6, use the following information. The members of the Professor of the Year Committee must pick a winner from three nominees: Tara Hammer (T), Mark Gerstein (H), and George Rodriguez-Paz (G). The preference table follows.

Number of Votes	5	3	2	2
First	T	H	H	T
Second	G	T	G	H
Third	H	G	T	G

1. How many members voted?

2. Does any candidate have a majority of votes?

3. Determine the winner using the plurality method.

4. Determine the winner using the Borda count method.

5. Determine the winner using the plurality with elimination method.

6. Determine the winner using the pairwise comparison method.

7. For a class project, Randi Beth will have her classmates vote for their favorite school subject. The choices are: Math (M), Science (S), English (E), and History (H). The results are given in the following preference table.

Number of Votes	10	8	7	5	5
First	M	H	E	E	M
Second	S	E	H	M	H
Third	E	S	M	H	S
Fourth	H	M	S	S	E

   a) What subject wins this election if the plurality method is used?
   b) What subject wins this election if the Borda count method is used?
   c) What subject wins this election if the plurality with elimination method is used?
   d) What subject wins this election if the pairwise comparison method is used?

8. Which voting method(s) – plurality, Borda count, plurality with elimination, or pairwise comparison – violates the head-to-head criterion using the following election data?

Number of Votes	50	30	20	15
First	A	C	A	B
Second	C	B	B	D
Third	D	D	D	A
Fourth	B	A	C	C

9. A group of 40 music students were asked to choose their favorite American Idol contestant from the top four contestants of the TV show for the 2002-2003 season. The choices were Reuben Studdard (R), Clay Aiken (C), Kimberly Locke (K), and Joshua Gracin (J). The 40 students ranked their choices according to the following preference table.

Number of Votes	18	10	8	4
First	C	R	K	J
Second	R	C	R	C
Third	K	K	J	R
Fourth	J	J	C	K

Using the data provided, if Joshua Gracin is eliminated, does the Borda counting method violate the irrelevant alternatives criterion?

10. In 2001, a community college hired 10 new adjuncts to teach in three different disciplines. The number of classes in each discipline is shown below.

Discipline	A	B	C	Total
Number of Classes	45	15	20	80

a) Find each discipline's apportionment using Hamilton's method.
b) Find each discipline's apportionment using Jefferson's method.
c) If the number of adjuncts hired that year increased to 11, does the Alabama paradox occur using Hamilton's method?
d) Suppose that in 2008 the disciplines have the following number of classes. Does the population paradox occur using Hamilton's method?

Discipline	A	B	C	Total
Number of Classes	52	18	25	95

e) Suppose that a fourth discipline and four additional adjuncts are added with the amount of classes shown below. Does the new-states paradox occur using Hamilton's method?

Discipline	A	B	C	D	Total
Number of Classes	45	15	20	28	108

A Survey of Mathematics with Applications, 7e

Chapter 15 – Voting and Apportionment   Form 2

In Exercise 1 – 6, use the following information. Employees of the Have A Nice Day office supply store voted for Employee of the Month. The three top vote-getters were Clay Jones (C), Danielle Amar (D), and Michele Bolger (M). The preference table follows.

Number of Votes	15	8	8
First	C	M	D
Second	D	C	C
Third	M	D	M

1. How many members voted?

2. Does any candidate have a majority of votes?

3. Determine the winner using the plurality method.

4. Determine the winner using the Borda count method.

5. Determine the winner using the plurality with elimination method.

6. Determine the winner using the pairwise comparison method.

7. At Appliances Plus, the manager took inventory of the top four appliances sold on Memorial Day. The four appliances were: a microwave (M), an outdoor gas grill (G), a toaster oven (T), and a refrigerator (R). The results are given in the following preference table.

Number of Votes	15	10	10	6
First	G	G	R	R
Second	M	M	T	T
Third	T	R	M	G
Fourth	R	T	G	M

a) Which appliance sold the most if the plurality method is used?
b) Which appliance sold the most if the Borda count method is used?
c) Which appliance sold the most if the plurality with elimination method is used?
d) Which appliance sold the most if the pairwise comparison method is used?

8. Which voting method(s) – plurality, Borda count, plurality with elimination, or pairwise comparison – violates the head-to-head criterion using the following election data?

Number of Votes	86	72	60	55
First	X	W	X	Y
Second	Z	X	W	X
Third	W	Y	Z	W
Fourth	Y	Z	Y	Z

9. A group of 60 RV owners were asked which U.S. cities were their favorites to visit. The top four choices were: Daytona Beach, Florida (D), Malibu Beach, California (M), Reno, Nevada (R), and Houston, Texas (H). The 60 RV owners ranked their choices according to the following preference table.

Number of Votes	22	18	11	9
First	D	M	D	R
Second	M	D	M	H
Third	R	H	H	M
Fourth	H	R	R	D

Using the data provided, if Malibu Beach is eliminated, does the Borda counting method violate the irrelevant alternatives criterion?

10. In 1998, a large fast food chain needed to apportion 50 new employees among three different stores based on the average number of customers each store has daily. This is shown below.

Store	A	B	C	Total
Number of Customers	550	620	430	1600

a) Find each store's apportionment using Hamilton's method.
b) Find each store's apportionment using Jefferson's method.
c) If the number of employees increased to 52, does the Alabama paradox occur using Hamilton's method?
d) Suppose that in 2005 the stores have the following average number of customers daily. Does the population paradox occur using Hamilton's method?

Store	A	B	C	Total
Number of Customers	675	600	714	1989

e) Suppose that a fourth store and ten additional employees are added. The number of average customers daily is shown below. Does the new-states paradox occur using Hamilton's method?

Store	A	B	C	D	Total
Number of Customers	550	620	430	515	2115

A Survey of Mathematics with Applications, 7e

In Exercise 1 – 6, use the following information. A new elementary school was just opened and the teachers of the school must pick a mascot from three choices: The Cobras (C), The Owls (O), and The Hurricanes (H). The preference table follows.

Number of Votes	8	7	5	2
First	C	C	H	O
Second	H	O	O	H
Third	O	H	C	C

1. How many members voted?

2. Does any candidate have a majority of votes?

3. Determine the winner using the plurality method.

4. Determine the winner using the Borda count method.

5. Determine the winner using the plurality with elimination method.

6. Determine the winner using the pairwise comparison method.

7. At the American Diamond Jewelry Exchange, the vendors were asked to choose their favorite gemstone. The choices were: Amethyst (A), Emerald (E), Sapphire (S), and Ruby (R). The results are given in the following preference table.

Number of Votes	15	10	9	9	6
First	E	E	R	A	R
Second	A	R	E	E	A
Third	S	S	A	S	S
Fourth	R	A	S	R	E

a) Which gemstone was the favorite if the plurality method is used?
b) Which gemstone was the favorite if the Borda count method is used?
c) Which gemstone was the favorite if the plurality with elimination method is used?
d) Which gemstone was the favorite if the pairwise comparison method is used?

8. Which voting method(s) – plurality, Borda count, plurality with elimination, or pairwise comparison – violates the head-to-head criterion using the following election data?

Number of Votes	50	50	45	10
First	A	A	D	B
Second	B	C	C	C
Third	C	D	B	D
Fourth	D	B	A	A

9. A group of 35 surfers were asked to choose their favorite brand of surfboard. The choices were: Surface (S), NSP (N), Softops (T), and Becker (B). The 35 surfers ranked their choices according to the following preference table.

Number of Votes	12	10	8	5
First	B	B	T	S
Second	T	T	N	N
Third	N	S	B	T
Fourth	S	N	S	B

Using the data provided, if Softops Surfboard is eliminated, does the Borda counting method violate the irrelevant alternatives criterion?

10. In 2004, Broward County hired 15 librarians to work in three new libraries. The distribution of librarians would be determined by the average number of items checked out from each library daily. The results are shown below.

Library	A	B	C	Total
Average Number of Items Checked Out Daily	200	350	170	720

a) Find each library's apportionment using Hamilton's method.
b) Find each library's apportionment using Jefferson's method.
c) If the number of librarians hired that year increased to 16, does the Alabama paradox occur using Hamilton's method?
d) Suppose that in 2006 the average daily number of items checked out is as follows. Does the population paradox occur using Hamilton's method?

Library	A	B	C	Total
Number of Items	250	370	200	820

e) Suppose that one library is added, and five additional librarians are hired. The average number of items checked out daily is shown below. Does the new-states paradox occur using Hamilton's method?

Library	A	B	C	D	Total
Number of Items	200	350	170	210	930

Chapter 1 Answers

Chapter 1 Form 1

1. 28, 33, 38

2. $\dfrac{1}{16}, -\dfrac{1}{32}, \dfrac{1}{64}$

3.
  a) The result is the original number times 3.
  b) The result is the original number times 3.
  c) The result will always be the original number times 3.
  d) $n, n+3, \dfrac{n+3}{2}, 6\left(\dfrac{n+3}{2}\right) = 3n+9$
     $3n+9-9 = 3n$

The answers for 4 – 6 are approximate.

4. 15,500,000

5. 18,000

6. 7 square units

7.
  a) ≈ 2.75 inches
  b) $\approx 6.5\%$

8. 25 checks

9. 21 folders

10. 4 hours

11. 11 times

12. Yes, Trevor is overpaid by $0.50.

13.

16	36	8
12	20	28
32	4	24

14. ≈ 0.05 hours would be gained.

15. 5, 21, 42, and 80

16. 18 hairclips

17.
  a) $120
  b) $119

  c) Larry would save $1 by using the coupon.

18. 24

Chapter 1 Form 2

1. 15, 19, 23

2. $\dfrac{16}{81}, \dfrac{32}{243}, \dfrac{64}{729}$

3.
  a) The result is the original number.
  b) The result is the original number.
  c) The result will always be the original number.
  d) $n, 6n, 6n+12, \dfrac{6n+12}{6}, = n+2$
  e) $n+2-2 = n$

The answers for 4 – 6 are approximate.

4. 522

5. 500,000,000

6. 11 square units

7.
  a) $450
  b) $150

8. 6 premium channels

9. 38 strawberries

10. $172\dfrac{1}{2}$ hours or 172 hours and 30 minutes.

11. 10 inches

12. Steve was not underpaid.

13.

11	21	7
9	13	17
19	5	15

14. Less time.

15. $2 \cdot 4 \cdot 6 \cdot 8 \cdot 12 = 4608$; 10 does not divide 4608.

16. 13 cars

17.
  a) $360
  b) $312
  c) Ming can save $48 by using the coupon.

18. 40,320

Chapter 1 Form 3

1. 24, 31, 38

2. $\dfrac{1}{256}, \dfrac{1}{1024}, \dfrac{1}{4096}$

3.
  a) The result is the original number times 2.
  b) The result is the original number times 2.
  c) The result will always be the original number times 2.
  d) $n, n-6, \dfrac{n-6}{2}, 4\left(\dfrac{n-6}{2}\right) = 2n-12$

  $2n - 12 + 12 = 2n$

The answers for 4 – 6 are approximate.

4. 60

5. 7,800,000

6. 6 square units

7.
  a) 1999 – 2000
  b) 25 cents per gallon

8. 221 minutes

9. 15 apples

10. 9 hours

11. 35 reruns

12. Yes, Robert was overpaid by $67.50.

13.

14	9	16
15	13	11
10	17	12

14. ≈ 0.02 hours would be saved.

15. 6, 11, 30, and 42

16. 64 marbles

17.
  a) $29.97
  b) $32.32
  c) Steven saves $2.35 if he buys the towels using the sale price.

18. 362,880

Chapter 2 Answers

Chapter 2 Form 1

1. False. $\{6\}$ is a subset, not an element.

2. True

3. False.
$\{J,\, S,\, R,\, K,\, T\}$ has 2^5 or 32 subsets.

4. True

5. False. $A \cap A' = \varnothing$

6. False. 2 is an even natural number.

7. True

8. False. The elements are not exactly the same in both sets.

9. True

10. $\{x \mid x \in N \text{ and } 4 \le x \le 8\}$

11. Set A is the set of natural numbers between 4 and 8, inclusive.

12. $\{0, 1, 3, 4, 5, 6, 7, 8\}$

13. $\{0, 8\}$

14. $\{1, 2\}$

15. 1

16.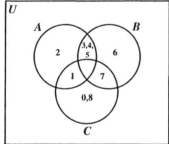

17. False

18.

U

green   blue

10   20   11

8

14   12

14

red

6

a) 10
b) 12
c) 14
d) 6

e) 54
f) 35

19. $\{4,\ 8, 12, 16, \ldots,\quad 4n,\ \ldots\}$
$\quad\ \downarrow \downarrow \downarrow \downarrow \qquad\quad\ \downarrow$
$\{8, 12, 16, 20, \ldots, 4n + 4, \ldots\}$

20. $\{1, 2,\ 3, 4,\ \ldots,\quad n,\ \ldots\}$
$\quad\ \downarrow \downarrow \downarrow \downarrow \qquad\quad \downarrow$
$\{2, 5, 8, 11, \ldots, 3n - 1, \ldots\}$

Chapter 2 Form 2

1. True

2. False. $\{5, 7\} = \{5, 7\}$ so it cannot be a proper subset.

3. True

4. False. $\{5\}$ is contained within the set.

5. False. S is an element, not a subset.

6. False. The set states $x < 8$, therefore 8 is not included in the set.

7. True

8. False. The two sets do not have the same number of elements.

9. False. The set is finite because it contains a specific number of elements.

10. $S = \{x \mid x \in N \text{ and } 1 \le x \le 4\}$

11. Set S is the set of natural numbers between 1 and 4, inclusive.

12. $\varnothing$

13. $\{2, 4, 6, 8, 10\}$

14. $\{2\}$

15. 6

16.

U

A   B

2   12

4,6   8,10

C

17. True

18.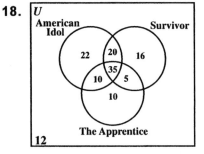

a) 48

b) 12

c) 35

d) 22

e) 5

f) 70

19. $\{2,\ 8,\ 14,\ 20,\ \ldots,\ 6n-4,\ldots\}$
$\quad\ \downarrow\downarrow\downarrow\downarrow\qquad\quad\ \downarrow$
$\{8,\ 14, 20, 26, \ldots,\ 6n+2,\ldots\}$

20. $\{1, 2, 3, 4,\ \ldots,\ n,\ \ldots\}$
$\quad\ \downarrow\downarrow\downarrow\downarrow\qquad\ \downarrow$
$\{2, 4, 6, 8,\ \ldots,\ 2n,\ldots\}$

Chapter 2 Form 3

1. True

2. False. *b* is not an element of the set.

3. False. The set has 2^4 or 16 subsets.

4. True

5. False. The sets are equal, therefore one set cannot be a proper subset of the other.

6. True

7. True

8. False. 4 is not inclusive, therefore it cannot be part of the subset.

9. True

10. $C = \{1, 2, 3, 4, 5\}$

11. Set *C* is the set of natural numbers less than 6.

12. $\{1, 10, 15, 20, 25\}$

13. $\{5, 15, 30\}$ or *C*

14. $\{1, 10, 20, 25\}$ or *B*

15. 2

16.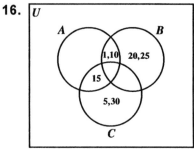

17. False

18.

football / baseball / hockey Venn diagram with values: 70, 20, 55, 40, 10, 15, 55, 15

a) 20

b) 45

c) 15

d) 55

e) 180

f) 10

19. $\{8, 10, 12, 14, \ldots, 2n+6, \ldots\}$
$\quad\ \downarrow\downarrow\downarrow\downarrow\qquad\quad\ \downarrow$
$\{10, 12, 14, 16, \ldots, 2n+8, \ldots\}$

20. $\{1, 2, 3, 4, \ldots,\quad n,\quad \ldots\}$
$\quad\ \downarrow\downarrow\downarrow\downarrow\qquad\quad\ \downarrow$
$\{5, 9, 13, 17, \ldots, 4n+1, \ldots\}$

Chapter 3 Answers

Chapter 3 Form 1

1. $(p \to q) \vee r$

2. $r \leftrightarrow (p \wedge q)$

3. $(q \wedge {\sim}r) \wedge p$

4. If Randi does not enjoy painting and Sierra does not take dance classes, then Jarred does not play baseball.

5. Jarred plays baseball and Sierra takes dance classes, or Randi does not enjoy painting.

6. T
   F
   F
   T
   T
   F
   T
   F

7. F
   T
   F
   F
   F
   T
   F
   T

8. False

9. True

10. True

11. False

12. Equivalent

13. (a) and (c) are equivalent.

14. (b) and (c) are equivalent.

15. $p \to q$
    $\underline{\quad q \quad}$
    $\therefore p$
    Invalid

16. Valid

17. Some dogs have tails.

18. No mosquitoes bite.

19. **Converse**: If I go shopping, then I will get paid today.
    Inverse: If I did not get paid today, then I will not go shopping.

Contrapositive: If I do not go shopping, then I did not get paid today.

20. Yes, if the conclusion does not follow from the set of premises.

Chapter 3 Form 2

1. $({\sim}r \vee p) \wedge {\sim}q$

2. $p \leftrightarrow ({\sim}q \wedge r)$

3. $(q \to {\sim}r) \vee {\sim}p$

4. Al does not follow the stock market or, if Tara does not love the computer then Jade reads magazines.

5. Tara loves the computer and, Al follows the stock market if and only if Jade does not read magazines.

6. T
   F
   T
   T
   T
   T
   T
   T

7. F
   T
   F
   F
   T
   F
   F
   F

8. False

9. True

10. False

11. True

12. Not equivalent

13. (a) and (c) are equivalent.

14. (b) and (c) are equivalent.

15. $p \vee q$
    $\underline{\quad {\sim}p \quad}$
    $\therefore q$
    Valid

16. Invalid

17. Some disco music is not great.

18. Consuelo does not like computers or Jacob does not like football.

19. **Converse**: If Amber will watch the Bachelorette, then she likes reality TV.
Inverse: If Amber does not like reality TV, then she will not watch the Bachelorette.
Contrapositive: If Amber will not watch the Bachelorette, then she does not like reality TV.

20. Yes, if the conclusion does not follow from the set of premises, the argument is invalid, even if the conclusion is a true statement.

Chapter 3 Form 3

1. $(p \wedge \sim q) \vee \sim r$

2. $r \rightarrow (\sim p \wedge \sim q)$

3. $(q \leftrightarrow \sim p) \wedge r$

4. If Sarah is an excellent seamstress or Irene is not a good cook, then Max does construction.

5. Irene is a good cook if and only if Max does not do construction, and Sarah is not an excellent seamstress.

6. T
F
F
F
T
T
T
T

7. T
T
T
T
T
T
T
F

8. True

9. False

10. True

11. True

12. Equivalent

13. (b) and (c) are equivalent.

14. (a) and (c) are equivalent.

15. $p \rightarrow q$
$\sim p$
$\overline{}$
$\therefore \sim q$
Invalid

16. Valid

17. No cats have green eyes.

18. Some babies are not cute.

19. **Converse**: If Suede's baseball team wins, then he was the pitcher.
Inverse: If Suede is not the pitcher, then his baseball team will not win.
Contrapositive: If Suede's baseball team does not win, then he was not the pitcher.

20. Construct truth tables for each statement. If both have the same truth values in the answer columns of the truth tables, the statements are equivalent.

Chapter 4 Answers

Chapter 4 Form 1

1. The Hindu-Arabic system
2. 20,305
3. 2954
4. 928
5. 5504
6. 4282
7. 1567
8. ſſ999∩∩∩|||||
9. MMMDCXLIII
10. 六百七十五
11. $\delta'\sigma o\eta$
12. (Mayan numeral)
13. In an additive system, the number represented by a particular set of numerals is the sum of the value of the numerals.
14. In a place-value system, each number is multiplied by a power of the base. The position of the numeral indicates the power of the base by which it is multiplied.
15. 23
16. 14
17. 11
18. 501
19. 11100_2
20. 830_9
21. 300301_4
22. 112440_5
23. 1021_6
24. 136_7
25. 1203_4
26. $220_5 \text{ R} 1_5$
27. 910
28. 3906

Chapter 4 Form 2

1. A system of numeration consists of a set of numerals and a scheme or rule for combining the numerals to represent numbers.
2. 1984
3. 5043
4. 862
5. 1082
6. 37,331
7. 102,460
8. MMMXXV
9. 八百七十
10. $\delta'\sigma\nu\varepsilon$
11. (Mayan numeral)
12. (Babylonian numeral)
13. In a ciphered system, the number represented by a particular set of numerals is the sum of the value of the numerals. There are numerals for each number up to and including the base and multiples of the base.
14. In a multiplicative system, there are numerals for each number less than the base and for powers of the base. Each numeral less than the base is multiplied by a numeral for the power of the base, and these products are added to obtain the number.
15. 59
16. 303
17. 21
18. 205
19. 201212_3

20. 14107_8

21. 101000000_2

22. 12221_7

23. 1102_3

24. 101_2

25. 1121_5

26. 330_4 R 1_4

27. 1520

28. 7930

Chapter 4 Form 3

1. The Babylonian system was not a true place-value system because it lacked a symbol for zero.

2. 3333

3. 12,070

4. 4675

5. 2596

6. 1303

7. 3904

8. $\alpha'\omega\lambda\eta$

9.

10.

11.

12.

13. In a place-value system, each number is multiplied by a power of the base. The position of the numeral indicates the power of the base by which it is multiplied.

14. In a ciphered system, the number represented by a particular set of numerals is the sum of the value of the numerals. There are numbers up to and including the base and multiples of the base.

15. 30

16. 663

17. 126

18. 209

19. 10051_6

20. 1100101100_2

21. 243141_5

22. 102332_4

23. 101000_2

24. 258_9

25. 11014_5

26. 342_8

27. 2325

28. 3675

Chapter 5 Answers

Chapter 5 Form 1

1. $2, 3, 4, 5, 6, 8,$ and 10

2. $2 \cdot 2 \cdot 2 \cdot 3 \cdot 31$

3. -3

4. -5

5. 54

6. $\dfrac{40}{11}$

7. $20\dfrac{1}{2}$

8. $0.\overline{1}$

9. $\dfrac{9}{400}$

10. $4\dfrac{103}{110}$

11. $\dfrac{7}{48}$

12. $5\sqrt{10}$

13. $3\sqrt{2}$

14. The natural numbers are not closed under the operation of subtraction. If a larger number is taken away from a smaller number, the result is an integer not a natural number.

15. The commutative property of addition.

16. The associative property of multiplication.

17. $\dfrac{1}{343}$

18. 4^5 or 1024

19. 3^8 or 6561

20. 6×10^{-10}

21. $a_n = 4n - 10$

22. $s_{13} = 52$

23. $a_7 = 20,480$

24. $s_7 = 889$

25. $a_n = 5 \cdot (-2)^{n-1}$

26. Yes. $2, 4$

Chapter 5 Form 2

1. $2, 3, 4, 5, 6, 8, 9$ and 10

2. $2 \cdot 2 \cdot 2 \cdot 5 \cdot 17$

3. -10

4. -8

5. 6

6. $\dfrac{70}{11}$

7. $20\dfrac{1}{6}$

8. $0.\overline{27}$

9. $\dfrac{63}{2000}$

10. $\dfrac{23}{126}$

11. $\dfrac{49}{60}$

12. $\sqrt{5}$

13. $\dfrac{\sqrt{15}}{5}$

14. Whole numbers are not closed under the operation of subtraction. If you subtract a larger number from a smaller number, the result is an integer.

15. The distributive property of multiplication over addition.

16. The commutative property of multiplication.

17. $\dfrac{16}{81}$

18. $\dfrac{1}{36}$

19. 7^3 or 343

20. 2.852×10^5

21. $a_n = 4n - 7$

22. $s_{10} = -265$

23. $a_7 = \dfrac{3}{32}$

24. $s_7 = 65$

25. $a_n = -2 \cdot (-3)^{n-1}$

26. Not a Fibonacci-type sequence.

Chapter 5 Form 3

1. $2, 3, 4, 5, 6, 8,$ and 10
2. $2 \cdot 2 \cdot 3 \cdot 3 \cdot 5 \cdot 5$
3. 3
4. 6
5. -28
6. $\dfrac{31}{6}$
7. $12\dfrac{1}{2}$
8. 0.125
9. $\dfrac{71}{2000}$
10. $5\dfrac{17}{36}$
11. $\dfrac{5}{72}$
12. $\sqrt{2}$
13. $\dfrac{\sqrt{10}}{2}$
14. The whole numbers are closed under the operation of addition. When two whole numbers are added together, the result is always a whole number.
15. The associative property of addition.
16. The commutative property of multiplication.
17. 2^6 or 64
18. $\dfrac{1}{36}$
19. 11^9 or $2,357,947,691$

20. 4×10^{11}

21. $a_n = -4n - 7$

22. $a_{12} = 978$

23. $a_6 = -2048$

24. $s_3 = 5\dfrac{7}{9}$ or $\dfrac{52}{9}$

25. $a_n = 12 \cdot \left(\dfrac{1}{2}\right)^{n-1}$

26. Yes. $-9, -15$

Chapter 6 Answers

Chapter 6 Form 1

1. 5

2. $\dfrac{4}{3}$

3. -1

4. $3x - 5 = 10;\ 5$

5. $x + 0.06x = 344.50;\ \$325$

6. 21

7. $y = x - 5$

8. 30

9. 22

10. $x < -2$

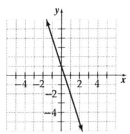

11. $-\dfrac{7}{6}$

12.

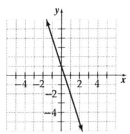

13.

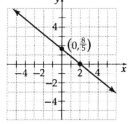

$\left(0, \tfrac{8}{5}\right)$

14.

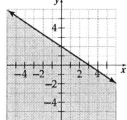

15. $-7, 8$

16. $\dfrac{-1 \pm \sqrt{113}}{8}$

17. Not a function. Does not pass the vertical line test.

18. -30

19.
 a) upward

 b) $x = 2$

 c) $(2, -1)$

 d) $(0, 3)$

 e) $(1, 0),\ (3, 0)$

 f)

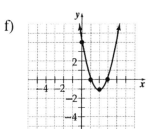

 g) Domain: $\mathbb{R}$
    Range: $y \geq -1$

Chapter 6 Form 2

1. -62

2. $\dfrac{13}{4}$

3. $-\dfrac{31}{11}$

4. $x + 6 = -12;\ -18$

5. $x + 0.20x = 25.60;\ \$32$

6. 31.5

7. $y = \dfrac{4x - 20}{5}$ or $y = \dfrac{4}{5}x - 4$

8. $\dfrac{80}{3}$

9. $\$285.71$

10. $x < 6$

11. $-\dfrac{3}{2}$

12.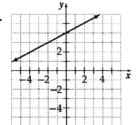

13.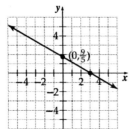

14.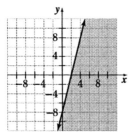

15. $-\dfrac{3}{2}, 5$

16. $\dfrac{-1 \pm \sqrt{97}}{8}$

17. Yes, it is a function. It passes the vertical line test.

18. -62

19.
 a) downward

 b) $x = 0$

 c) $(0, 4)$

 d) $(0, 4)$

 e) $(-2, 0), (2, 0)$

 f)

g) Domain: $\mathbb{R}$
   Range: $y \le 4$

Chapter 6 Form 3

 1. 108

 2. $\dfrac{1}{2}$

 3. 4

 4. $2x - 7 = 21;\ 14$

 5. $2(w + 3) + 2w = 103;$
  $W = 24\dfrac{1}{4},\ L = 27\dfrac{1}{4}$

 6. 3360

 7. $y = \dfrac{-2x + 15}{5}$ or $y = -\dfrac{2}{5}x + 3$

 8. 360

 9. 3 hours

 10. $x > -2$

 11. 4

 12.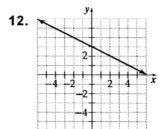

 13.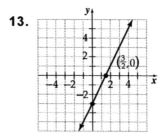

 14.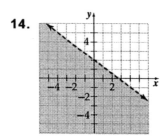

15. $-4, 5$

16. $\dfrac{-3 \pm \sqrt{65}}{4}$

17. Not a function. Does not pass the vertical line test.

18. -10

19.

a) upward

b) $x = 3$

c) $(3, -1)$

d) $(0, 8)$

e) $(2, 0), (4, 0)$

f)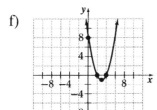

g) Domain: $\mathbb{R}$
Range: $y \geq -1$

Chapter 7 Answers

Chapter 7 Form 1

1. The graph would be two parallel lines.

2.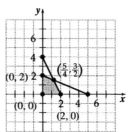

3. An infinite number of solutions.

4. $(3, 3)$

5. $(0, -1)$

6. $\left(0, -\dfrac{5}{4}\right)$

7. $(7, 3)$

8. $(5, -3)$

9. $(3, -2)$

10. $\begin{bmatrix} -7 & 7 \\ 11 & 3 \end{bmatrix}$

11. $\begin{bmatrix} -18 & 13 \\ 19 & 2 \end{bmatrix}$

12. $\begin{bmatrix} -30 & 7 \\ 7 & -20 \end{bmatrix}$

13.

14. 5 pounds of dark chocolate; 10 pounds of milk chocolate

15.
a) 31 miles
b) Ride-Around-Town would be more expensive.

16.
a)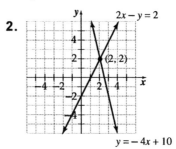

b) minimum profit: $(0, 0)$

   maximum profit: $\left(\dfrac{5}{4}, \dfrac{3}{2}\right)$

Chapter 7 Form 2

1. The graph would look like it contained one line because both lines would be in the same exact place on the coordinate system.

2.

3. No solution

4. $(3, -1)$

5. $(-4, -3)$

6. $(3, 4)$

7. $(2, -3)$

8. $(7, 4)$

9. $(2, -1)$

10. $\begin{bmatrix} -3 & 3 \\ -1 & -2 \end{bmatrix}$

11. $\begin{bmatrix} 21 & -1 \\ -1 & 14 \end{bmatrix}$

12. $\begin{bmatrix} -18 & -5 \\ 6 & -9 \end{bmatrix}$

13.

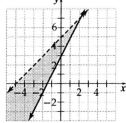

14. 25 child tickets;
15 adult tickets

15.
a) 20 cars
b) used cards

16.
a)

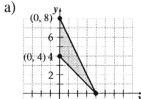

b) minimum profit: $(0, 4)$
maximum profit: $(0, 8)$

Chapter 7 Form 3

1. Two lines would intersect at only one point.

2.

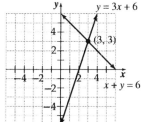

3. Exactly one solution.

4. $(1, 2)$

5. $(2, 0)$

6. $(-2, 9)$

7. $(15, 12)$

8. $(3, 3)$

9. $(2, 3)$

10. $\begin{bmatrix} -10 & 8 \\ -5 & -2 \end{bmatrix}$

11. $\begin{bmatrix} 9 & -22 \\ -7 & 6 \end{bmatrix}$

12. $\begin{bmatrix} -20 & 16 \\ -10 & -4 \end{bmatrix}$

13.

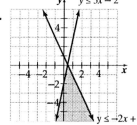

14. 0.5 gallons of 100% anti-freeze;
2 gallons of 30% anti-freeze

15.
a) 3 bags of plant food
b) The Town and Country Nursery

16.
a)

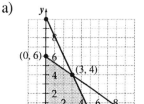

b) minimum profit: $(0, 0)$
maximum profit: $(3, 4)$

Chapter 8 Answers

Chapter 8 Form 1

1. 0.165 m
2. 22,500 dg
3. 1000 times greater
4. 4 kg
5. (b)
6. (c)
7. (a)
8. (b)
9. (b)
10. 1,000,000 times greater
11. 1000 times greater
12. 49.5 kg
13. ≈ 246 in.
14. ≈ −6.7° C
15. 23° F
16. 300 cm
17.
  a) 60,000 cm³
  b) 60 ℓ
  c) 60 kg
18. $1173

Chapter 8 Form 2

1. 35,000 dg
2. 1.28 ℓ
3. 10 times greater
4. 5.4 m
5. (a)
6. (b)
7. (c)
8. (b)
9. (c)
10. 1,000,000 times greater
11. 1,000,000,000 times greater
12. 58.5 m
13. 17 ft
14. ≈ 237° C

15. 35.6° F
16. 2.4 m
17.
  a) 28,080 cm³
  b) 28.08 ℓ
  c) 28.08 kg
18. $125

Chapter 8 Form 3

1. 0.056 dam
2. 430 da ℓ
3. 1000 times greater
4. 10 ℓ
5. (c)
6. (b)
7. (b)
8. (c)
9. (b)
10. 1,000,000 times greater
11. 1,000,000,000,000,000 times greater
12. 5.4 m
13. 29.5 oz
14. ≈ −21° C
15. 53.6° F
16. 4 yd
17.
  a) 240,000 cm³
  b) 240 ℓ
  c) 240 kg
18. $51

Chapter 9 Answers

Chapter 9 Form 1

1. $\overset{\circ\quad\circ}{\overline{AC}}$

2. $\measuredangle ABC$

3. $\overline{BC}$

4. $\{C\}$

5. $47.2°$

6. $101.8°$

7. $123°$

8. $1260°$

9. $56\,\text{cm}$

10.
   a) $20\,\text{cm}$
   b) $48\,\text{cm}$
   c) $48\,\text{cm}^2$

11. $\approx 628.3\,\text{ft}^3$

12. $61{,}250\,\text{cm}^3$

13. $\approx 904.8\,\text{in}^3$

14.
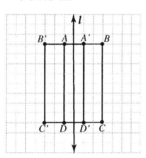

15.
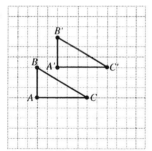

16.
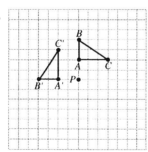

17.
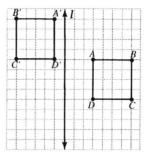

18.
   a) Yes
   b) No

19. A Klein bottle is a topological object that resembles a bottle but has only one side.

20. Answers/sketches will vary.
   a) No holes
   b) One hole

21. Given a line and a point not on the line, one and only one line can be drawn parallel to the given line through the given point.

Chapter 9 Form 2

1. $\measuredangle ABC$

2. $\overline{CD}$

3. $\overrightarrow{BD}$

4. $\triangle ABD$

5. $27.9°$

6. $5.5°$

7. $125°$

8. $720°$

9. $16\,\text{in.}$

10.
   a) $12\,\text{ft}$
   b) $36\,\text{ft}$
   c) $54\,\text{ft}^2$

11. $\approx 212\,\text{cm}^3$

12. $\approx 226\,\text{in}^3$

13. $84\,\text{m}^3$

14.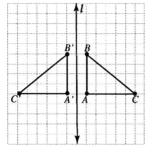

15.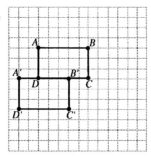

16.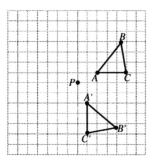

17.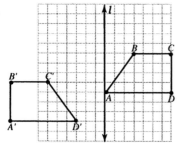

18.
   a) No
   b) Yes

19. A Jordan curve is a topological object that can be thought of as a circle twisted out of shape.

20. Answers/sketches will vary.
   a) 2 holes
   b) 3 or more holes

21. Given a line and a point not on the line, no line can be drawn through the given point parallel to the given line.

Chapter 9 Form 3

1. $\{A\}$

2. $\sphericalangle ABC$

3. $\overline{BC}$

4. $\{\ \}$

5. 81.6°

6. 80.1°

7. 69°

8. 900°

9. $\approx$ 31 m

10.
   a) 8 ft
   b) 24 ft
   c) 24 ft^2

11. 1331 ft^3

12. 1.2 m^3

13. 132 in^3

14.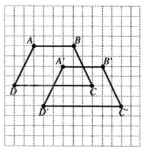

15.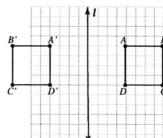

16.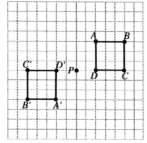

17.

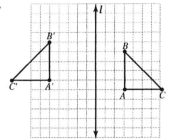

18.
 a) No
 b) Yes

19. Take a strip of paper, give one end a half twist, and tape the ends together.

20. Answers/sketches will vary.
 a) No holes
 b) 2 holes

21. Given a line and a point not on the line, two or more lines can be drawn through the given point parallel to the given line.

Chapter 10 Answers

Chapter 10 Form 1

1. A binary operation is an operation or rule that can be performed on two and only two elements of a set.

2. Yes, for a group the commutative property need not apply.

3. No, positive integers are not closed under the given operation.

4.

+	1	2	3	4
1	2	3	4	1
2	3	4	1	2
3	4	1	2	3
4	1	2	3	4

5. Yes, it is a commutative group.

6. 4

7. 1

8.
   a) ✥
   b) Yes
   c) Yes, A
   d) R
   e) T

9. Yes, it is a commutative group.

10. No, there is no identity element.

11. Yes, it is a commutative group.

12. 1

13. 2

14. 1

15. 6

16. 3

17. 7

18. 3

19. 4

20.
   a)

X	0	1	2	3
0	0	0	0	0
1	0	1	2	3
2	0	2	0	2
3	0	3	2	1

   b) No, no inverse for 0 or 2.

Chapter 10 Form 2

1. A counterexample is a specific example illustrating that a specific property is not true.

2. No, every commutative group is also a group.

3. Yes, multiplying integers will always yield an integer.

4.

+	1	2	3	4	5	6
1	2	3	4	5	6	1
2	3	4	5	6	1	2
3	4	5	6	1	2	3
4	5	6	1	2	3	4
5	6	1	2	3	4	5
6	1	2	3	4	5	6

5. Yes, it is a commutative group.

6. 3

7. 1

8.
   a) ✿
   b) Yes
   c) Yes, C
   d) A
   e) Y

9. No, it is not a commutative group.

10. Yes, it is a commutative group.

11. Yes, it is a commutative group.

12. 1

13. 2

14. 3

15. 5

16. 7

17. No solution.

18. 3

19. 2

20.

a)

X	0	1	2	3	4	5
0	0	0	0	0	0	0
1	0	1	2	3	4	5
2	0	2	4	0	2	4
3	0	3	0	3	0	3
4	0	4	0	0	4	2
5	0	5	4	3	2	1

b) No, not every element has an inverse.

Chapter 10 Form 3

1. Another name for a commutative group is an abelian group.

2. If a binary operation is performed on any two elements of a set and the result is an element of the set, then the set is closed under the given binary operation. For all integers a and b, $a + b$ is an integer.

3. No, the quotient of two natural numbers may yield a rational number which is not always a natural number.

4.

+	1	2	3
1	2	3	1
2	3	1	2
3	1	2	3

5. Yes, it is a commutative group.

6. 1

7. 1

8.

a) ✳

b) Yes

c) Yes, H

d) B

e) B

9. Yes, it is a commutative group.

10. No, it is not commutative.

11. Yes, it is a commutative group.

12. 1

13. 5

14. 1

15. 1

16. 2

17. 3

18. 1

19. 2

20.

a)

X	0	1	2
0	0	0	0
1	0	1	2
2	0	2	1

b) No

Chapter 11 Answers

Chapter 11 Form 1

1. $40
2. 60%
3. $360
4. $6760
5. $4702.86
6. $202.86
7. $8370.66
8. $3790
9. $1125
10. $33
11. 5.5%
12.
  a) $39.47
  b) $1978.91
13.
  a) 5.0%
  b) $77.56
  c) $3084.43
14.
  a) $8.51
  b) $546.83
  c) $461.80
  d) $6.93
  e) $468.73
15. $75,000
16. $7493.33
17. $2098.13
18. $1470
19. $1882.50
20. Yes
21.
  a) $516,000
  b) $266,000

Chapter 11 Form 2

1. $31.40
2. 6 months
3. $206.25
4. $2706.25

5. $3854.18
6. $354.18
7. $7773/79
8. $1983.59
9. $308.75
10. $15.25
11. 4.5%
12.
  a) $128.21
  b) $1309.29
13.
  a) 7.0%
  b) $150.50
  c) $4269.50
14.
  a) $3.44
  b) $1327.19
  c) $1092.55
  d) $13.11
  e) $1105.66
15. $26,250
16. $6960.83
17. $1949.03
18. $1291.15
19. $1520.32
20. Yes
21.
  a) $309,876
  b) $134,876

Chapter 11 Form 3

1. $14.55
2. $5100
3. $293.25
4. $6043.25
5. $4818.42
6. $318.42
7. $5689.81
8. $15,369.91
9. $2960

10. $190

11. 8.0%

12.
  a) $176.13
  b) $1578.87

13.
  a) 7.5%
  b) $222.85
  c) $3906.99

14.
  a) $17.50
  b) $999.90
  c) $1066.89
  d) $21.38
  e) $1088.27

15. $108,500

16. $8441.67

17. $2363.67

18. $1408.49

19. $1891.82

20. Yes

21.
  a) $507,056.40
  b) $197,056.40

Chapter 12 Answers

Chapter 12 Form 1

1. $\dfrac{2}{5}$

2. $\dfrac{5}{11}$

3. $\dfrac{5}{11}$

4. $\dfrac{9}{11}$

5. $\dfrac{3}{11}$

6. $\dfrac{2}{11}$

7. $\dfrac{3}{11}$

8. $\dfrac{2}{11}$

9. $\dfrac{2}{11}$

10. $\dfrac{4}{13}$

11. 8

12.

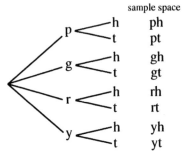

13. $\dfrac{1}{8}$

14. $\dfrac{5}{8}$

15. $\dfrac{3}{8}$

16. 1,051,044,800

17.
 a) 4:3
 b) 4:3

18. $\dfrac{2}{9}$

19. $1.83

20.
 a) $\dfrac{25}{63}$

 b) $\dfrac{26}{63}$

 c) $\dfrac{4}{13}$

 d) $\dfrac{3}{10}$

21. 720

22. $\dfrac{1}{22}$

23. $\dfrac{21}{22}$

24. $\dfrac{175}{663}$

25. 0.2936

Chapter 12 Form 2

1. $\dfrac{1}{4}$

2. $\dfrac{2}{5}$

3. $\dfrac{1}{2}$

4. $\dfrac{9}{10}$

5. $\dfrac{1}{5}$

6. $\dfrac{3}{25}$

7. $\dfrac{2}{15}$

8. $\dfrac{2}{15}$

9. $\dfrac{1}{3}$

10. $\frac{8}{13}$

11. 12

12.

```
                          sample space
              1    H1
              2    H2
        H     3    H3
              4    H4
              5    H5
              6    H6

              1    T1
              2    T2
        T     3    T3
              4    T4
              5    T5
              6    T6
```

13. $\frac{1}{4}$

14. $\frac{3}{4}$

15. $\frac{1}{6}$

16. 6,500,000

17.
a) 1:4
b) 1:1

18. $\frac{5}{7}$

19. $2.46

20.
a) $\frac{81}{160}$

b) $\frac{23}{80}$

c) $\frac{29}{46}$

d) $\frac{32}{57}$

21. 1320

22. $\frac{2}{21}$

23. $\frac{19}{21}$

24. $\frac{35}{143}$

25. 0.0244

Chapter 12 Form 3

1. $\frac{1}{3}$

2. $\frac{4}{5}$

3. $\frac{7}{15}$

4. $\frac{4}{15}$

5. $\frac{2}{3}$

6. $\frac{1}{5}$

7. $\frac{11}{21}$

8. $\frac{4}{15}$

9. $\frac{1}{6}$

10. $\frac{7}{13}$

11. 6

12.

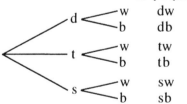

```
                          sample space
              w    dw
        d     b    db

              w    tw
        t     b    tb

              w    sw
        s     b    sb
```

13. $\frac{1}{6}$

14. $\frac{2}{3}$

15. $\frac{1}{6}$

16. 9,000,000

17.
a) 8:7
b) 8:7

18. $\dfrac{5}{9}$

19. $1.25

20.

 a) $\dfrac{23}{40}$

 b) $\dfrac{11}{20}$

 c) $\dfrac{7}{17}$

 d) $\dfrac{5}{11}$

21. 154,440

22. $\dfrac{6}{11}$

23. $\dfrac{5}{11}$

24. $\dfrac{97{,}020}{323{,}323} \approx 0.3001$

25. 0.003

Chapter 13 Answers

Chapter 13 Form 1

1. 14.2
2. 15
3. 20
4. 13
5. 14
6. ≈ 6.18
7.

Class	Frequency
10 – 15	4
16 – 21	2
22 – 27	2
28 – 33	3
34 – 39	4
40 – 45	3
46 – 51	2

8.

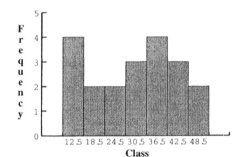

9.

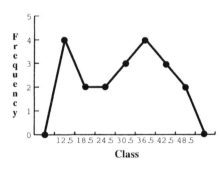

10. 150
11. 175
12. 75%
13. 88%
14. 8300
15. 232
16. 133
17. 6.1%
18. 95.3%

19. 25.1%
20. 153 houses
21.
a)

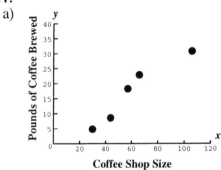

b) Yes
c) ≈ 0.963
d) Yes
e) $y = 0.35x - 4.01$
f) 29.24

Chapter 13 Form 2

1. ≈ 14.2
2. 12
3. 6
4. 15
5. 26
6. ≈ 0.09
7.

Class	Frequency
25 – 30	5
31 – 36	6
37 – 42	5
43 – 48	4

8.

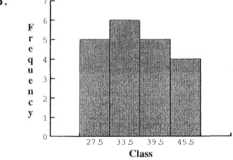

9.

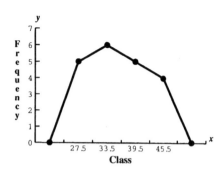

10. 485

11. 470

12. 75%

13. 53%

14. 3000

15. 380

16. 560

17. 53.3%

18. 93.3%

19. 6.7%

20. 171

21.

a)

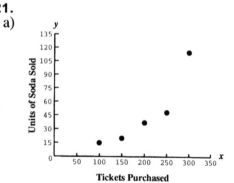

b) Yes

c) ≈ 0.9

d) No

e) $y = 0.456x - 44.2$

f) ≈ 36

Chapter 13 Form 3

1. 10

2. 10

3. 7

4. 10.5

5. 7

6. ≈ 3.08

7.

Class	Frequency
50 – 59	4
60 – 69	3
70 – 79	2
80 – 89	6
90 – 99	5

8.

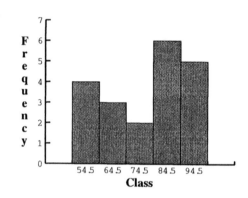

9.

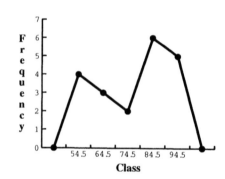

10. 850

11. 1000

12. 75%

13. 90%

14. 196,000

15. 1055

16. 930

17. 14.7%

18. 77.3%

19. 93.3%

20. ≈ 176

21.

a)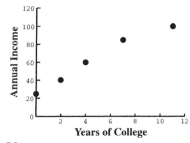

b) Yes

c) ≈ 0.985

d) Yes

e) $y = 7.05x + 28.16$

f) $91.61 thousand or $91,610

Chapter 14 Answers

Chapter 14 Form 1

1. Graphs will vary, such as:

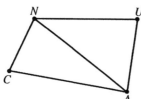

2.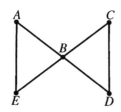

3. Graphs will vary, such as:

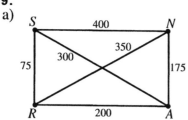

4. Answers may vary, such as:
   BAEDCAFE

5. No

6. Answers may vary, such as:
   ABHCIBCDFEGDEFGHIJA

7. Answers may vary, such as:
   ABCDEFGA

8. 6! = 720 ways

9.
a)

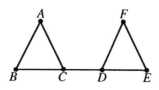

b) *N,A,R,S,N* or *N,S,R,A,N* for $850
c) *N,A,R,S,N* for $850

10. Graphs may vary, such as:

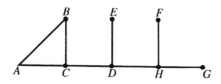

11.

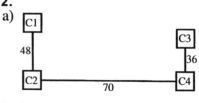

12.
a)

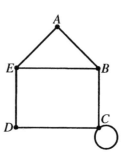

b) $115.50

Chapter 14 Form 2

1. Graphs will vary, such as:

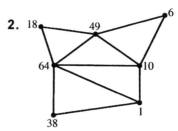

2.

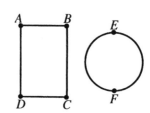

3. Graphs will vary, such as:

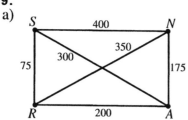

4. Answers may vary, such as:
   ABCBABC

5. Yes. The person may start in any room, and the person will end in the room he or she started in.

6. Answers may vary, such as:
ADCBDCBA

7. Answers may vary, such as:
ABCDEFA

8. 4! = 24 ways

9.

a)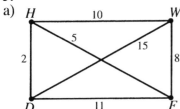

b) *H,F,W,D,H* or *H,D,W,F,H* for 30 miles

c) *H,D,F,W,H* for 30 miles

10. Graphs may vary, such as:

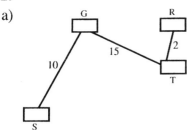

11.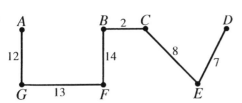

12.

a)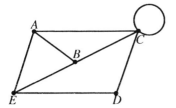

b) $149.85

Chapter 14 Form 3

1. Graphs will vary, such as:

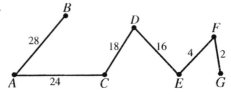

2.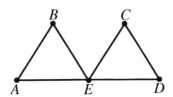

3. Graphs will vary, such as:

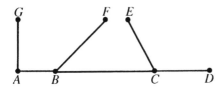

4. Answers may vary, such as:
ADCBABCDB

5. Yes. The person may start in any room, and the person will end in the same room where he or she started.

6. Answers may vary, such as:
EABCDECADE

7. Answers may vary, such as:
ABCDEFGHA

8. 4! = 24 ways

9.

a)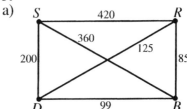

b) *S,R,B,D,S* or *S,D,B,R,S* for $724
c) *S,D,B,R,S* for $724

10. Graphs may vary, such as:

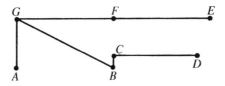

11.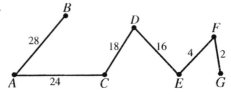

12.

a)

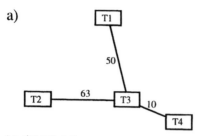

b) $338.25

Chapter 15 Answers

Chapter 15 Form 1

1. 12
2. Yes
3. Tara Hammer
4. Tara Hammer
5. Elimination is not necessary since Tara Hammer received the majority of votes.
6. Tara Hammer
7.
   a) Math
   b) Math
   c) English
   d) English
8. Plurality with elimination
9. Yes
10.
   a) 6, 2, 2
   b) 6, 2, 2
   c) No
   d) Yes
   e) No

Chapter 15 Form 2

1. 31
2. No
3. Clay Jones
4. Clay Jones
5. Clay Jones
6. Clay Jones
7.
   a) The outdoor gas grill
   b) The outdoor gas grill
   c) The outdoor gas grill
   d) A three way tie between the microwave, the refrigerator, and the outdoor gas grill.
8. None of the voting methods violates the head-to-head criterion.
9. No

10.
   a) 17, 19, 14
   b) 17, 20, 13
   c) No
   d) No
   e) Yes

Chapter 15 Form 3

1. 22
2. Yes
3. The Cobras
4. The Cobras
5. Elimination is not necessary since The Cobras received the majority of votes.
6. The Cobras
7.
   a) Emerald
   b) Emerald
   c) Emerald
   d) Emerald
8. None of the voting methods violates the head-to-head criterion.
9. No
10.
   a) 4, 7, 4
   b) 4, 8, 3
   c) No
   d) No
   e) No